CLASSIC and Vintage COMMERCIALS

on

ERF

Published by
KELSEY PUBLISHING LTD

Printed in Singapore by Stamford Press PTE Ltd.
on behalf of
Kelsey Publishing Ltd,
Cudham Tithe Barn,
Berry's Hill,
Cudham,
Kent TN16 3AG
Tel: 01959 541444
Fax: 01959 541400
E-mail: kelseybooks@kelsey.co.uk
Website: www.kelsey.co.uk

ISBN: 1 873098 68 5

INTRODUCTION

A decade ago, ERF could proudly boast that it was the only surviving British lorry manufacturer, its lorries being built at the legendary Sun Works in Sandbach, Cheshire - the same factory from in which the first example of the marque had left the production line in 1933. They still benefited from the enviable reputation for strength and reliability built up over 70 years and the words 'ERF - Built in Britain' were proudly displayed on the side of each vehicle.

Today, ERF is a subsidiary of German manufacturer MAN, and British manufacture has ceased. Though you can still buy an ERF-badged lorry, it is basically a MAN with a few different badges and a revised front grille.

It is not only have recent times that have been turbulent for ERF, because the company originally stemmed from a family dispute. Foden had been struggling to sell in renowned Steam Waggons and Edwin Richard Foden, youngest son of the company's founder, was convinced the future lay in diesel lorries.

A huge company row over this led to ER Foden and his son Dennis to resign and set up their own company called ER Foden, later to become known as ERF. The company was radical in many ways, not least the emphasis on diesel lorries when most other manufacturers favoured petrol. The policy from the start was to rely on the best bought in components available rather than making everything in-house and this philosophy would serve ERF well throughout its independent existence.

From the beginning, ERF's company's products were widely sought after in both home and export markets, justifying its one time advertising claim to be 'The World's best oil-engined lorry. ERF innovations over the years ranged from the panoramic windscreened early 1950s KV cab, the unique (if slow selling) semi forward control Sabrina and, for the 1970s, the revolutionary SP (Steel Plastic) cab. ERF had few failures in more than six decades of independent production and the quality and reputation of its products meant the company was able to survive the downturns which had seen the demise of so many British lorry manufacturers, though the early 1980s were undoubtedly a difficult time.

Prosperity returned and sales increased massively, so many were surprised when ERF was sold to Western Star of Canada in 1996 and, in 2000, when MAN of Germany took over. This was heralded as a new beginning with millions of pounds invested in a new factory at Middlewich, celebrated with a major 'Coming Home – Moving Home' event in August 2000 when more than 200 ERFs gathered at Sun Works for a cavalcade to the new facilities.

ERF's CEO John Bryant described the occasion as marking '"the second most important stage in ERF's history." No need to comment further apart from the fact that British ERF manufacture ceased in 2002, a year before the marque's 70th anniversary.

Cheshire-built ERFs are still a common sight in service and 1980s examples are still the backbone of many fairground fleets.

Nick Larkin
Editor, *Classic & Vintage Commercials*
March 2004

CONTENTS

Friend of the ERF

Nicholas Watts is a farmer by profession but he is also a keen ornithologist and amateur historian. His fascination for the past has led him to create his own museum of bygone agricultural implements and artefacts.

By Peter Davies

Coupled with this he has always been fascinated by heavy lorries. One make in particular has a special appeal to 52-year-old Nick and that is ERF. However, as a lad, it was Leylands that caught his imagination.

"When I was about eight I used to be impressed by the Leyland eight-wheelers of Walter Smith, the potato merchants from Gosberton," he explains. "They would come to my father's farm here at Deeping St Nicholas and my interest sprang from there. Lorries have fascinated me ever since."

Nick has recently rebuilt a 1936 ERF CI6.82 which began life with Somerset haulier S J Cox of Street, but his first restoration job began nearly two decades ago with a similar ERF of 1938 vintage.

It was in October 1977 that Nick decided to look for an ERF eight-wheeler to restore. At that time no one had rescued an ERF of that type, in fact eight wheelers generally were a somewhat neglected species.

He put a 'Wanted' ad in *World's Fair* and received a reply from Day's Amusements from Port Talbot in South Wales. Day's did not pretend that the lorry was in good condition but their caution that the brakes would need attention turned out to be the understatement of the year.

"I already owned an early Foden recovery vehicle, a 1932 R-type," says Nick, "so I set off in that down to Port Talbot. When I got there I discovered tha the old ERF was completely derelict - the rust had even eaten through the rear end the chassis frame."

The truck had stood for 14 years near the sea front and there are few things worse than salt air for corroding steel. Th cab, too, was in a very dilapidated state. I spite of this Nick decided to buy the ERF which was registered in Lancashire as

'TC 308. It was duly hitched up to the
oden and taken home on a suspended
w.
Back at his farm Nick and his part-time
elper set about the long task of renovating
TC which, at that time, was by far the
dest example in captivity. "We stripped it
ght down to the chassis frame. I've never
en anything in such a state," says Nick.
decided to get the whole thing
otblasted by a local firm at Langtoft Fen

new cab and paint job. The ERF, the first such model to enter preservation, has been fitted with an overdrive box and higher ratio diff to give it extra road speed. Nick always makes a point of driving it to rallies rather than taking it piggy-back. It was originally geared to 30mph maximum which, in this day and age, is a bit too slow.

While DTC was still undergoing improvements Nick got thinking about another ERF eight-wheeler. "I saw an advert in the HCVS Newsletter in 1985," he recalls. "It referred to a 1936 eight wheeler for sale by a dealer in Ripon, Yorkshire. I thought that as I had DTC 308, which at that time still had its non standard cab, I might as well get the other one."

There wasn't much left of the one at

so we could see the full extent of the corrosion."

It was an enormous task but Nick and his helper stuck at it throughout the following year, and by the end of 1978, the chassis was making good progress. Several visits were made to scrapyards in search of missing parts. Nick even acquired a set of period style 36x8 tyres which would add authenticity to the finished job.

When Nick first got DTC he obtained a build sheet from ERF at Sandbach and began finding out more about the 40-year-old truck's history. It turned out to have begun life with quarry owners Keirby & Perry of Blackpool. In its early days it suffered a couple of accidents, one of which was fairly major because it had necessitated the replacement of both chassis sidemembers.

"The replacements were 11in deep," says Nick, "while the original ones were only 9in. This resulted in the cab and body sitting higher than they should with large clearances under the mudguards."

The cab which was on DTC when Nick first bought it was a peculiar hybrid affair bearing a distinct resemblance that of a post-war Maudslay. It was decided to restore it in this form but, after a few appearances at rallies, he realised it was worth looking for a proper ERF cab. Not surprisingly there wasn't one to be found in servicable condition so a replica was built, based on a derelict original found at a large Stevenage scrapyard.

At the same time Nick decided to recreate the original owner's colour scheme. The only guide was found in latter day colour photos of Keirby & Perry's tipper fleet in the fifties so the livery was based on these. It later transpired that some of K & P's platform lorries may have been painted blue but the situation is somewhat unclear.

Regardless of the exact livery DTC looked the part when it re-emerged with its

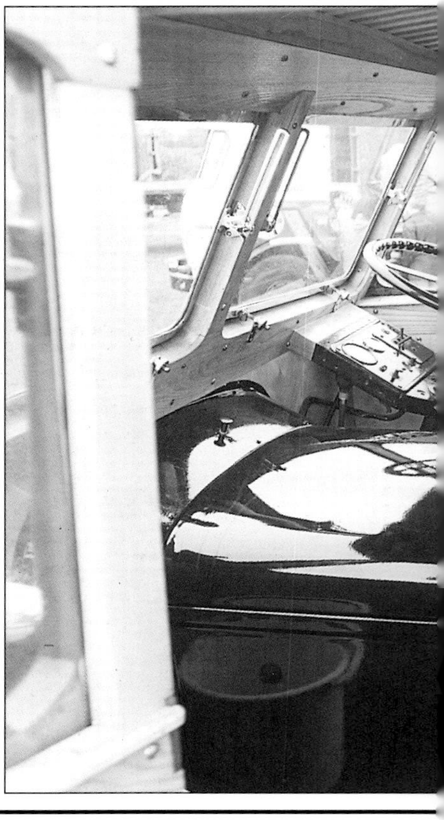

ɾipon but there turned out to be a third CI6.82 which had once been operated by Elliott of York. This had been cut down to a fifth wheel tractor but was still driveable. Nick decided to have both. Only one log book existed and that was for the remains of the Cox vehicle. This motley mixture of ERF remains lay in Nick's barn until 1990 before restoration began.

The cab on the Elliott tractor wasn't worth restoring because the joints were coming apart and the timber was quite rotten. So Nick approached David Templeman, an expert craftsman in building wooden frame car bodies for Rolls-Royce and Alvis. Having built an exact replica timber frame, this was then taken to specialist panel beater, Ron Cantrell, at Stoke-on-Trent.

Ron used to skin cabs at ERF. In the old days it used to cost about £3.50 to skin one completely but nowadays it costs 'an arm and a leg.' Ron also made the distinctive ERF style wings which are exact copies of the originals.

The whole process of restoring BYB 375 took five years and Nick, who is busy looking after the family farms, employed several people to assist him with the task. The lorry has the very early David Brown

O4 four-speed direct crash gearbox and a top speed that is scarcely higher than that of a bicycle. Nick decided to fit it with 40x8 wheels and tyres which at least gave it a touch more speed.

When it came to the paint job he decided it would look good in a typical pre-war dark blue with black chassis and has had it signwritten with his own name. The Donington-based coach painter who did the spray job was none other than the one who used to spray the Walter Smith Octopuses back in the fifties.

BYB had quite a chequered history. Being an early example it is based on a medium wheelbase with a 22ft 6in body

and the original operators regretted their choice because the body was too short. They had another CI68.2 on the long wheelbase with a 24ft body and that was much more practical. They used to carry a lot of bagged cement and BYB always had to be loaded with an extra layer to bring it up to its full 22-ton gross.

In 1939, after just three years' work, BYB caught fire one day and was severely damaged. The cement bags burned away and the chassis was engulfed in cement. It was left to stand in Cox's yard for some time and eventually the engine was taken out and put into the other ERF eight-wheeler when that broke down with engine failure.

Soon after that the burnt-out engineless remains of BYB were taken in by ERF for refurbishment. The vehicle's history then becomes a little obscure but S J Cox became Cox Mid Somerset Transport and were eventually absorbed into British Road Services. The lorry is believed to have been involved in a serious accident in 1952 when it ran off the road near Tewkesbury. After that it was written-off.

When not dabbling with ERFs Nick divides his spare time between bird watching, shooting and building up his growing local history museum. The prize exhibit is a very old sail reaper built in Thirsk.

Nick farms 1,500 acres of arable land spread across seven family-owned farms around Deeping St Nicholas in the Fens. Crops include wheat, oil seed rape, peas, linseed, beet and potatoes. He is in the process of restoring a unique horse-drawn railway which apparently was once a common feature of Fenland farms in bygone days.

The 2ft gauge railway would have been one and a half miles long and was used between 1920 and 1939 to haul produce from distant fields up to the farm buildings. It was de-commissioned in the fifties and fell into disrepair but traces of it still remain. Nick's enthusiasm for bygones apparently has no bounds.

BLOOD SWEAT AND TEARS

"Merlin" talks to truck salesman Jim Neary about his Welsh ERF.

Head south from Oswestry, Newton and Llanidoes, or north from Builth Wells and Rhyader on the A470 and you will eventually arrive at a roundabout whose westerly direction on the A44 will point you towards Aberystwyth across the Plynlimon mountain range.

It is a journey to be enjoyed during the brighter months of the year but not so pleasant when the nights have closed in. Travel the road for 12 or so miles and eventually you will pass through the small village of Ponterwyd. Or will you?

My journey terminated on the outskirts of the village, at Minafon to be precise with a renovated cottage on one side and an ERF A series tipper parked directly opposite on a small but adequate hardstanding. The cottage and the contents of the small yard are owned by Jim and Lynne Neary. "The house is hers and the wagon is mine," is my initial greeting from Jim Neary who emerges from under GDE645N, armed with grease gun at the ready.

The Welsh A series ERF just as Jim Neary purchased the vehicle in March 1991. Photo Alan Spillet.

GDE645N is a smartly renovated 6x4 tipper with aluminium triple dropside body (maker unknown) and Edbro tipping gear. Propulsion for this 21ft-bodied beauty comes from the legendary Gardner 180 6LXB six-cylinder 10,450cc diesel engine, or is that "oiler" I can hear you say? The motor is original bar the replacement exhaust manifold and her number is 194946. The big power unit of its day transmits raucous energy via the David Brown six-speed box which drives the GVW'd 24.390 tonner through two

The completed project - the ERF A series looking resplendent just ready to set off for a local rally.

Halfway home, photo shot in Aberaeron, August 1992. On the left we have ace mechanic Gary Munford and on the right Jim Neary. Photo Alan Spillet.

'he broken rear exhaust mounting is clearly visible in our picture.

Tanks for the memory". An early shot of the restoration shows chassis and bodywork efore restoration had begun.

ɔot the difference - GDE645N looking resplendent at the Welshpool Festival of Transport the summer of 1995. Martin Platt.

The Leyland fuel tank was promptly disposed of to make way for an appropriate ERF vessel. Note chassis and outrigger refurbishment. the new side raves and mud guards are in place. Photo Jim Neary.

The cab under going the final stages of rebuilding.

Kirkstall axles. Bearing chassis number 29554, this particular vehicle was first registered on March 1 1975. The A series model range entered the haulage market in 1970. With the arrival of the new B series in mid-1975, it is prudent to speculate that this was one of the last LV-cabbed models to enter service in 6x4 form.

Jim bought the tipper in March 1991 from a haulier in Pembrokeshire through a chance sighting while on business.

The first owner was S J Talbot, a fruit and vegetable dealer from Merlin's Bridge near Haverfordwest. Apart from local duties, it regularly ran to Covent Garden for bulk collection of goods. From Talbot it passed to John Lawrence of Clarebeston Road. While operating for the two companies GDE ran as a flatbed. Curiously, the original ERF data declares the vehicle as having a tipper chassis. Jim surmises that it is possible the chassis was the result of an cancelled order and, being

After removing the dash ABS warning unit a plate of spaghetti appeared so the A series had to have its wiring upgraded.

readily available, was consequently flat-bodied for Talbot's requirements, although this has yet to be substantiated.

After service with Lawrence, the lorry passed into the hands of Hubert Day & Sons, also of Haverfordwest. It was here that the flat was removed and replaced by the current tipper body.

When the clocking off card was punched at Day's the tired old lady found a new home at Ray Rees & Sons of Robeston, Wathen. Sadly, years of hard graft and

Taking a rest as the cab is shown in the final stages of restoration.

Jim Neary stands proudly outside the cab with GDE645N, awaiting a call up to pick up a load.

Jim heads through the centre of Aberystwyth town centre in February 1995; the hard hat on the nearside of the cab is for authenticity.

modern technology had caught up with GDE. A great deal of mechanical refurbishment was carried out by the Rees family, mainly with regard to the working parts of the chassis and suspension, plus a radiator. It was operated on short haul work for a brief time but its age and general deterioration finally led to a reluctance by the workforce to drive it as more comfortable and powerful vehicles became available.

"I paid a big one for it," Jim says with a smile. "Basically it was very good mechanically, thanks to Ray and the boys. The cab was well shot and the body was tatty, but it fired up pretty well considering that it stood for several months over the winter. My friend Gary checked it over and I bought it."

Having duly parted with hard-earned cash and a firm handshake, GDE graciously left the Pembrokeshire yard and headed north for Cardiganshire. Recounting the first trip Jim declared mixed feelings of delight and tradition. "It was a lousy evening, Gary drove and I listened. I seemed to hear many strange noises." (Don't we all!). The lorry ran well all the way, about 30 miles to the first resting place. A brief interruption occurred, however, when the silencer fell off about one mile further down the line. This was duly rectified by wiring up the remains of the then rear-mounted box to the chassis. There was no immediate remedy for the noise, though.

While it was parked safely off the road in Aberaeron for the next 18 months, basic repairs were gradually carried out. These included the fitting of a decent set of tyres, new batteries, exhaust system and an engine service. August 1992 saw the return of GDE to the tarmac as once more it set off for the 31-mile journey to its current home at Ponterwyd.

Safely home at last and across the road from the cottage, Jim could now finally devote more time to preservation work. "My initial target was to finish in time for the ERF 60 years of success at Oulton Park on September 11 1993," he relates. "It was the event I was really looking forward to." Did he make it? "Negative," came the reply. "I was held back by finances, time and bad weather."

He did eventually finish it, 14 months o in November 1994, with a full Ministry ticket in February 1995 and a local rally in May of that year. Jim proudly proclaims a smoke test reading of 0.4 for his Gardner, "It made my day!" he said cheerfully.

All in all a great deal of time, effort and dedication have been put into this preservation project. Apart from the engin and main cab structure the whole vehicle has been renovated. It looks good and a decent run up the road will prove to anybody that GDE runs as sweet as a nut, even though she's 21 years old next March A final comment from Jim: "It was blood, sweat and tears."

As I climb into my car ready for the journey home, I realise there was one question I hadn't asked. "What do you do for a living, then?" I said. "I sell trucks," he retorted with a five-mile grin.

We thank Alan Spillett and Jim Neary fo their help in this article; we look forward t "Merlin's" next rendition which will be on a Welsh transport operator who goes back many years.

ERF Sold!

ERF Holdings, the UK's last remaining manufacturer of heavy commercial vehicles, has agreed a sale to a Canadian group for £27.4 million.

The Cheshire based company which was founded in 1933 and is celebrating 63 years of business is the subject of a recommended cash offer at 275p a share from Western Star Trucks Holdings based in British Columbia, Canada.

A Western Star truck working hard in North Vancouver docks last September, caught on camera by the Editor.

Another Western Star with Caterpillar power which is disconnected from its oversize lowloader trailer. Seen at Edmonton, Alberta, Canada.

It took the driver a considerable amount of skill to reverse this outfit down to the hotel restaurant in Jasper, Alberta, Canada. The bonnet is all fibreglass and is typical of most US-style manufactures. This machine carries a Caterpillar engine and runs on 24in rims, the wheel studs are quite small.

Last month Western Star bought 4.2 per cent of ERF's shares and now accounts for 52.4 per cent of the company. The company has now obtained the agreement of the chairman Peter Foden to sell the 30 per cent stake which he owns together with family trusts. Institutions M & G and Newton Investment Management have also undertaken to sell holdings of 12.5 per cent and 5 per cent respectively.

This offer represents a premium of 55 per cent over ERF's share price of six weeks ago and a 13 per cent premium over the price on May 16. This was the day before the company disclosed it was in discussions with Western Star.

Peter Foden said Western Star was an excellent fit for ERF, which employs 800 staff . "We are strong in the UK and Africa and are making inroads into Europe while Western Star is very strong in Canada, the US, Australia, Far East and the Middle East."

Terry Peabody, Western Star chairman who looked lean and fit, said there were synergies in engineering research, purchasing and marketing. Terry Peabody is highly resepected for his skill in turning companies into profit as he did with Western Star starting in 1991, who are just about the same size as ERF. Most of Western Star's customers are owner-drivers or small operators and they have 1.12 per cent of the US truck market; in 1995 they sold 2,250 trucks.

ERFs pre-tax profits dipped from £2.3 million to £1.8 million in the year to the end of March. This was despite being boosted by an exceptional profit of £800,000 from the sale of 70 per cent of its South African subsidiary. A final dividend, payable on August 15, maintains the total at 4.5p.

So the last UK commercial vehicle manufacturer goes to the other side of the water. Let's hope we see the company go from strength to strength and we see no redundancies, but jobs created! We will be watching developments with interest.

Western Star's history goes back through White Trucks of Cleveland, Ohio who are now mixed up with Volvo. Western Star is a popular truck particularly in west of the Canada and the US. The Western Star trucks look to be built on very traditional lines, many are fitted with Caterpillar and Cummins engines, and they are made at Kelowna, British Columbia.

As ON THE WESTCOUNTRY TOBERS

Anderton & Rowland

[redacted]

Above, DAW280L is still in use with the No. 1 Dodgems and is pictured pulling on for Penzance Corpus Christi Fair on 5 June 1994. The loads need to pull over on to the opposite side of the road in order to negotiate the narrow entrance to the recreation ground where the fair is held.

The firm of Anderton & Rowland is one of the country's most well-known family of amusement caterers. For several generations, the company has been operated by the De Vey family who are based at Cullompton, Devon and Bristol.

Perhaps better known for the immaculate fleet of Scammell Showtracs that they once operated (hopefully the subject for a future issue), the ERF 'A' series has nevertheless seen significant use with the firm during the past 25 years.

First to join the vehicle fleet was an 8-legger GBA158C which was acquired in

GBA158C pictured at Blackwater, Cornwall in May 1980, en route to Redruth Whit Fair. This lorry operated with the firm's No. 1 Dodgems.

The supreme showman's livery. It may be raining but GYA827K "Ark Royal" still looks immaculate as it hauls the No. 2 Dodgem loads along the Redruth Bypass en route to the town's Whit Fair on 23 May 1993.

Also nearing Penzance on 6 June 1993, OTT220M "Lord Mountbatten" with arcade and staff trailer in tow.

he early 1970s by the late Ernie De Vey, nr to carry the platforms, nets and cenery of the firm's No. 1 Dodgems. itted with frame bodywork and originally ravelled in white livery, this was later hanged to the more traditional maroon. he remained in service until the end of e 1986 season when replaced by a 'B' eries.

For 1984, the venerable Scammell howtrac "John Bull" was taken out of ervice with the No. 1 Dodgems, now perated by Ernie De Vey, Jnr, and eplaced with an 'A'series tractor)AW280L, purchased from the Somerset-ased haulage firm of Oswall. Still in ervice today, carrying twin generators and owing a box truck and paybox, she has to ate seen 13 years of continuous use.

For the commencement of the 1981 eason, the transport for the firm's No. 2)odgems, operated by George De Vey, vas revamped with the complete ride being carried on one 8-wheeler and box truck. The new lorry, an 'A' series GYA827K was operated in plain maroon livery for the first year but during the next few seasons was to be lettered and lined out to become a superb travelling advertisement. It has generally become tradition for the A&R vehicles to be given names. This one, which is still in regular use, is known as "Ark Royal". In addition to the large box truck, the ride's paybox is also often hitched on at the rear.

At the end of the 1981 season, another of the famous Showtracs, "Dragon", for many years associated with the Speedway Ark, was also put into honourable retirement. The replacement vehicle was a 6-wheeled 'A' series tractor with extended chassis NBX956H, which was given the name "Lord Nelson", a title handed down from the days of steam. Fitted with box

NBX956H "The Lord Nelson" negotiates a roundabout with its train as it arrives at Penzance on 6 June 1993.

body which houses no fewer than three generators, it is still regularly used and tows two living trailers. The operator is Colin De Vey who travels two rides with the No. 1 section.

Continued next month.

A^s ON THE WESTCO
TOBERS

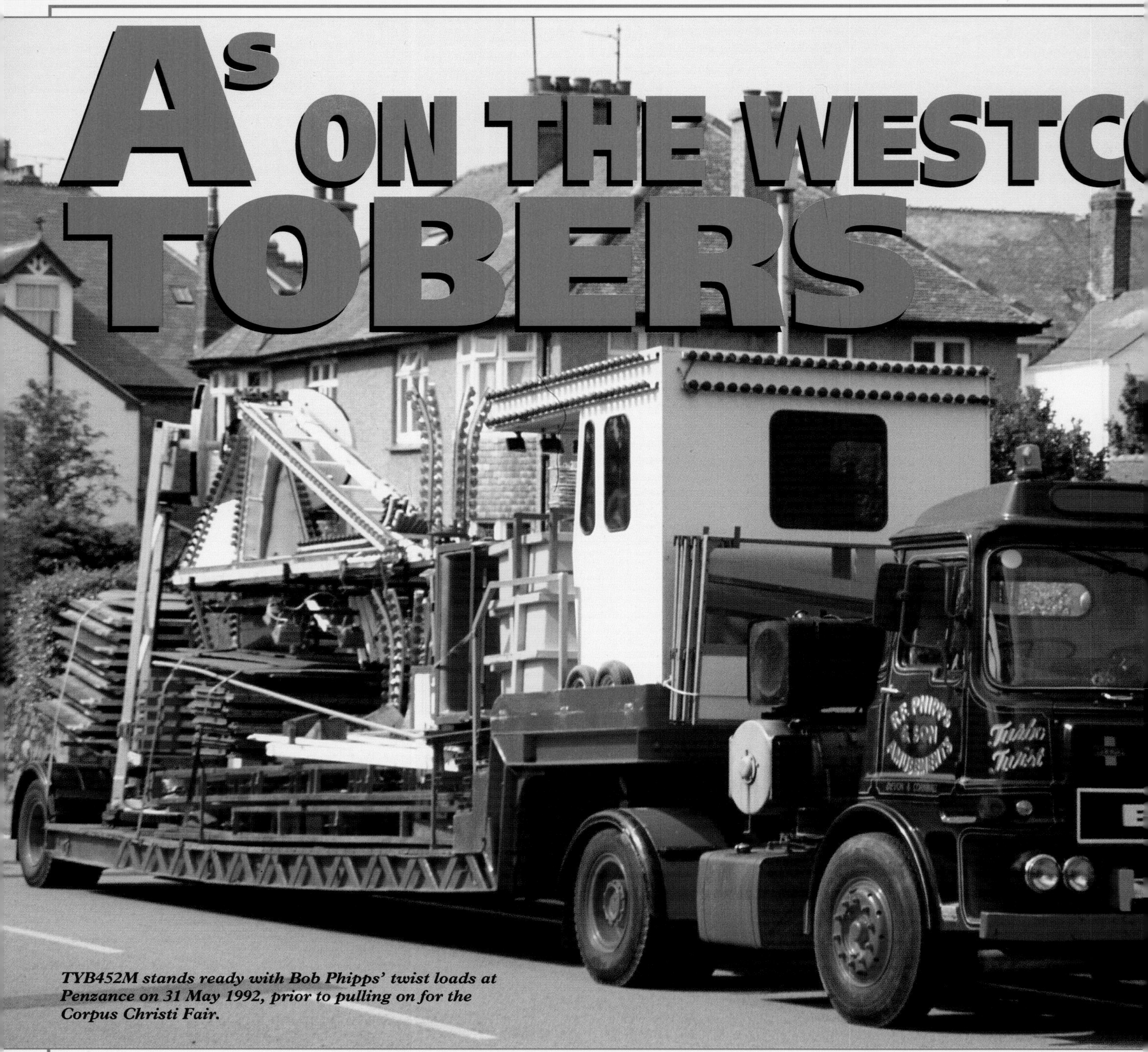

TYB452M stands ready with Bob Phipps' twist loads at Penzance on 31 May 1992, prior to pulling on for the Corpus Christi Fair.

We continue a look at the firm Anderton & Rowland and its well-known operators the De Vey family, who have catered for the amusements industry for several generations.

In 1984, Simon De Vey (son of George) also commissioned a 6-wheeled 'A' series OTT220M which he christened 'Lord Mountbatten'. Carrying three generators and initially hauling his Lifting Paratrooper load, it was later used to pull his arcade and a staff trailer, and was traditionally finished in the firm's standard livery. Following mechanical failure, it was withdrawn during the 1994 season and replaced with a 'B' series which was to bear the same name. Though out of use, it is still retained by the firm.

Michael De Vey, Ernie's brother, travels the Whitelegg circuit with his twist ride

Newly pressed into operation to haul Michael De Vey's twist loads, XYD899N still in its former owners' livery (Oswall) stands alongside the vehicle it deposed, AHY716J 'Trader Horn', in its final year of service. Redruth 20 May 1984.

NTRY

d slip, as well as joining the firm for any of its traditional venues. He has erated no fewer than three 'A' series hicles since the mid-1970's. The first is a tractor unit AHY716J, (replaced an eing Scammell), and carried two nerators, towing the Twist loads roughout Devon and Cornwall. It bore e name 'Trader Horn' (originally used an FWD operated by A&R during the 30's). First travelled in black, it was er turned out in blue livery, and in 1984 is relegated to lighter duties for a season the acquisition of an articulated 'A' ries tractor, which was put to use as insport for the twist (previously the ride d been fitted with a dolly, the artic spensed with the need for this and so cut wn on travelling time). This new lorry, YD899N, also came from Oswall ansport and was still in their livery when essed into use during the first season. bsequently finished in blue and named

XYD899N, in new livery as 'American Express' overtakes a later 'B' series OYA791P 'Montgomery' (No. 2 Dodgems) as it arrives with the twist loads for Helston Flora Fair, 3 May 1992.

WTU202H ascends 'The Big Dipper' along the Camborne-Hayle Bypass on 24 April 1988 as it makes its way to Penzance for the annual Spring Fair.

"American Express", it travelled until the end of the 1992 season. The third 'A' was an 8-wheeler WTU202J acquired and operated during the 1980s as a replacement for 'Trader Horn' and used on doublers to haul Michael's living trailer and kiosk. Both vehicles have now been replaced by a 'B' and a 'C' series.

Bob Phipps, who sadly passed away during 1995, married George De Vey's sister, and travelled the A&R circuit for many years. A Canadian by birth, he adapted readily to the English showman's way of life, and after operating as a tenant, became a riding master in his own right, firstly with an Octopus and later a Twist. A Scammell was used with this latter ride during the first season, and thereafter, from 1984, an 'A' series was used which also came from Oswall's. This artic unit TYB452M is still in use as the ride's prime mover, with Bob's son, Robert, having now taken the reins as riding master. A year previous to the acquisition of this vehicle another 'A' had been brought into service, purchased in June 1993 from Cornish haulier, Sid Knowles of Mabe, near Falmouth. JAF456L carried two generators and was originally operated with open back. This was later boxed in and the unit ran doublers towing its owner's living trailers and arcade. It was withdrawn at the end of the 1993 season, and it is understood that following breaking up, some components were exported to the Far East.

To complete this chapter, it is appropriate to include one of A&R's most regular tenants, the late Tommy Opie, who operated out of Bristol with his two juvenile rides, a traditional platform set and an Austin car ride. For many years, the Austin 7s were transported on an 'A' series 6-wheeler box van, HMO224N. The lorry is in fact still in use, as the rides have been acquired by Anderton Bros of Crediton, Devon who still travel them, attending a number of the A&R fairs, as well as numerous preservation events and one day festivals.

To complete the picture, the next episode will deal with the smaller operators of Devon and Cornwall, both riding masters and tenants.

Tommy Opie arrives at Truro on 30 May 1989 with the Austin Car Ride aboard HMO224N in readiness for the city's annual Whit Fair.

Also seen climbing, 'The Big Dipper' on 28 May 1991, Bob Phipps en route for Hayle Fair with JAF456L towing the living trailers.

HILL START

By Peter Davies

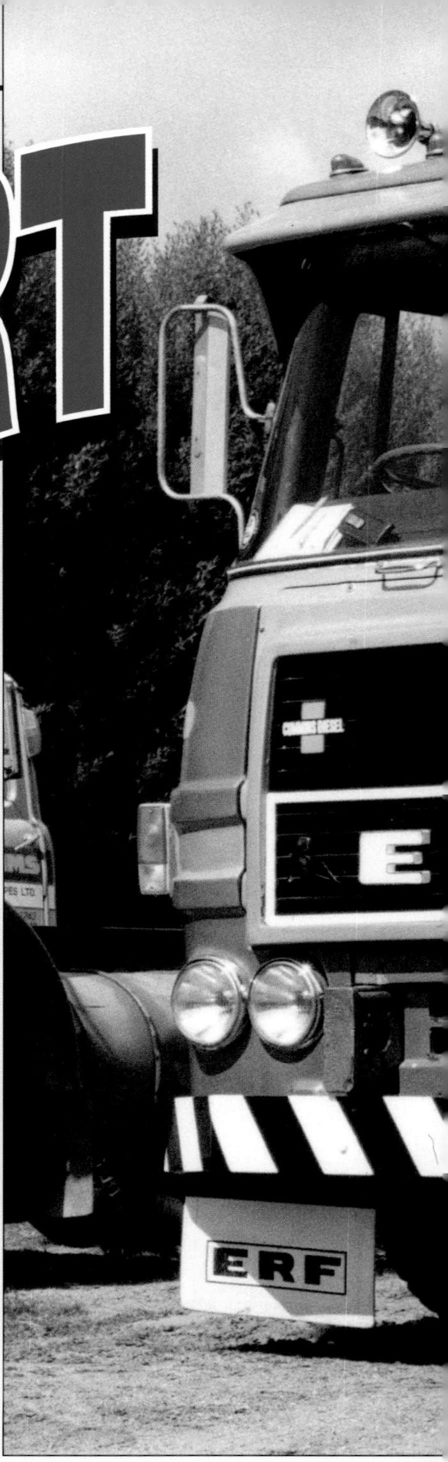

Once upon a time ERF A-series, or 'A-lines', as they are often called in New Zealand, were a common enough sight, but now they have moved into the realms of preserved trucks. There are quite a few already preserved by British enthusiasts but 'down under' they're fast becoming an endangered species.

Eventually they might all have gone to the great 'junkyard in the sky', were it not for the gallant efforts of one avid ERF fan based in Rotorua - Kerry Hill, who is the appointed New Zealand representative of the UK Register of ERF Vehicles Society (REVS for short), and has recently fulfilled his ambition of owning one of the Sandbach classics.

During a recent visit to witness the latest trends in Kiwi trucking, a number of us 'Poms' called on Kerry, where we were let in on a closely guarded secret. Although it was common knowledge that he had already acquired a 1967 F21OD International last year, no one was expecting to be confronted by an ERF A-series!

Truck preservation in New Zealand is not nearly as widespread as it is in the UK. There are several reasons for this. Perhaps the main one is that there are virtually no concessions for the would-be preservationist. You might think us Poms are hard done by, but at least we can use old vehicles on reduced, or even zero tax rates. If you want to own an old truck in New Zealand, even though it is not for commercial use, you must pay the full rates of road user charges (that's the equivalent of our Vehicle Excise Duty) as well as keeping a log of every journey. To make life even more difficult, all trucks have to go through a twice-yearly fitness test.

On top of this, there is no nationally-based club or society to cater for the truck enthusiast, whereas in the UK we have at least three such clubs as well as several dedicated one-make clubs like REVS, the AEC Society, the Foden Society and the like.

Perhaps sheer numbers might have something to do with it. New Zealand is relatively sparsely populated and there are correspondingly fewer trucks. And there is another major factor - namely high costs. The average second-hand truck is at least two or three times the price it would be in the UK. Consequently, operators keep them longer and the average enthusiast can't justify such a large outlay.

The A-series dates from 1975 and is powered by a 250 Cummins.

New Zealand trucks survive better thanks to the sunnier climate, and operators keep them working longer owing to high replacement costs.

WL3328 is a newly-issued registration number, but the rest of the truck is very original.

In spite of all these drawbacks, Kerry ha taken the plunge and saved one of the fast-disappearing A-series tractor units. It is now safely tucked away in the garage at hi family home in Rotorua. The 'Inter' has been passed on to another enthusiast, so that too will be preserved.

Kerry got to hear about the old ERF through a mate who drives for Dynes Transport of Tapanui on South Island. It was at Ettrick, which lies on Highway 8 between Milton and Alexandra. He first found out about it a couple of years ago, but the owner didn't want to sell it. Renewed attempts to agree a deal were made in September 1997. The asking pric of around £2,000 wasn't too bad by New Zealand standards, but it was unrestored

Kerry Hill has taken the initiative of preserving a classic ERF in New Zealand where there are no concessions for individual enthusiasts.

Kerry's A-series needed a lot of renovation to bring it up to scratch.

ıd needed quite a lot of work to bring it
p to scratch.

One of the sticking points in negotiating
ıe deal was over who should cover the
ɔst of registration and the CoF
Certificate of Fitness) - and of a new hub
ıeter. The truck had stood in a shed from
991 to 1997 and had consequently lost its
riginal registration. Before it could be
riven on the public road it had to undergo
ıe statutory six-monthly CoF test and be
sued with new registration number and
ıe regulation hub meter, which registers
ıe number of kilometres clocked to
etermine the amount of RUC (Road User
harges) the owner has to pay. Kerry
ısisted that the cost of all this (about
250) should be borne by the vendor.
ventually the matter was settled after
ɔout three weeks of haggling.

Having agreed a deal it was decided to
rive the truck back to Rotorua, which meant a three hour ferry crossing from Picton to Wellington. Not surprisingly, the road journey had its anxious moments, as one might expect with a truck that has been standing for six years.

While en route to Kaikura, the radiator core split and began losing water - about 20 litres were lost and Kerry was in the middle of nowhere without the necessary tools to sort out the problem. Answer? Plug the hole with chewing gum - yes, chewing gum! At least it enabled him and his mate to get to Kaikura, where they bought some 'stopper' (Radweld equivalent) and topped up the system, which was dangerously low by then.

There was also a leak in the diesel tank, and they were losing diesel all the way home - the leak had not shown that much until the tank was filled up for the journey. All in all, the ERF was in a bit of a rough state and, according to Kerry, the batteries on it had completely had it (although he used stronger words!). "The truck was ticking over when we arrived to collect it," he explains. "We made the mistake of stopping the motor when we stopped for a break and there wasn't enough in the batteries to re-start the engine. We had to buy two new batteries from a local garage in order to get going again." When Kerry dipped the engine oil, that too was well down, and took eight litres (one and three quarter gallons!) to bring it up to the full mark.

It is surprising that it passed its CoF as it was, as there were quite a few jobs needed to get it up to scratch. Once Kerry had got it safely back to Rotorua he could take stock and tidy the old girl up. A new set of 'boots' was a must, so he opted for Hankook 10.00R20s, which have a standard highway tread pattern reminiscent of the old Michelins. It was on

Farr air cleaner and intake stack has been remade to pattern.

Fibreglass cab has survived well for 23 years.

9.00x20s before, so the bigger tyres gave it a bit more speed, if only 10kph. Where it used to do 80kph (50mph) at 1900 revs it now does 90kph.

Before the ERF could venture out on the road again it needed a new fuel tank, as the old one was corroded and leaking. Kerry has now fitted a very smart polished alloy tank, giving the old girl a real facelift. Another job which was a must was the air cleaner. Once again, corrosion had set in, and Kerry ended up having a new one fabricated to the same pattern. It was originally supplied by GKN Farr of Birmingham. Both Farr and the more familiar Donaldson Cyclopac filters were fitted by ERF, depending on the model. A new vertical air intake stack in alloy cost £100. Inevitably, the exhaust was clapped, so that too has been completely replaced with a new system. The support bracket for this has been modified to make it stronger.

The cab and the mechanical units were not in too bad a shape. When the light blue and grey paint job was completed, the old A-series began to look as good as new. Finishing touches included new mudflaps all round and some discreet signwriting. Now it is 'number one' in the K D Hill fleet - let's hope that means there's more in the pipeline.

Kerry's ERF has quite an interesting history. It was first built on August 4, 1975 and is Chassis No 29269. It is powered by a Cummins NHC250 and has a 'round the houses' overdrive 9-speed Fuller Roadranger box. Rear axle is a Kirkstall hub reduction with a ratio of 5.6:1. When new it went to Talley's Fisheries at Port Motueka on South Island, and was delivered on August 28, 1975. There it pulled a 25ft tandem fridge trailer built by Roadrunner of Vaughan Road, Rotorua.

On July 6, 1979 she was traded in to ERF dealers ISP (Industrial Steel & Plant) for a new B-series 4x2 unit. ISP sold it on to Mosgiel-based Tairei Carrying Co in March 1980 and was put back on reefer work, hauling meat all over central South Island for a year or so.

Tairei sold the truck to ERF dealer Domtrac Equipment Ltd at Dunedin in May 1981, where it spent a couple of years as a spare tractor unit for 'back up' use when another rig was off the road. The last owners, oddly enough, went under the name of Hill (but no relation to Kerry). They were M G Hill & Sons, Contractors, based at Millers Flat, Ettrick near Roxburgh. It was there that Kerry found her and finally bought her for preservation on October 2, 1997.

After thorough research, Kerry has managed to come up with photographs of the ERF with all its previous owners. The truck has been re-registered twice, and now carries a 1997 number WL3328.

Last-known operators of the A-series were M G Hill of Ettrick, where Kerry tracked her down in 1997.

Originally Kerry thinks it was 'HH????' b he is still trying to check this out. While a Tairei Carrying Co, it had been re-registered as JI7528.

Good news is that the original trailer ha turned up at Talley's in Motueka, and Kerry hopes to buy this soon and put the rig back to original. Getting it back home is a possible problem, as it is not in a roa worthy condition and is too tall to low lo it on a trailer on the ferry. Until he manages to resolve the problem, he is thinking of putting a 2-ton ballast weight over the fifth wheel to make the rig a bit more smooth-riding.

So many old trucks in New Zealand ge run into the ground or end up being give the hot spanner treatment. There are too few keen enthusiasts like Kerry who are prepared to save them and, let's face it, th system is all geared against the preservationist. Kerry has not only restor the old A-series, he has also made a comprehensive survey of all the ERFs he can trace in New Zealand, travelling man miles to photograph them and record the chassis numbers.

As such, he has contributed a lot of valuable information to REVS back in th UK. Let's hope enough 'Kiwis' can be found to form a club to promote the preservation of a few more trucks before they end up being scrapped. Things like the unique ERF MW and MV rigid eight used to be popular, but now they are disappearing. It would be nice to think th at least one or two such motors could be saved for posterity.

Kerry's truck was first owned by Talley's Fisheries at Motueka.

Hard-worked ERF is seen here, in service with Taierei Carrying Co, with whom it worke with a reefer, carrying meat.

E
IN SOUTH S

By Ale

This August Bank Holiday I decided to venture a little further afield, spending a day at Weston Park, host to the South Staffordshire Town & Country Fayre. One of the main attractions was the display of many ERF lorries in the 'pERFectionists' show. This well established and successful event has built up in size and reputation over the years and offers the whole family an excellent programme of entertainment, including helicopter rides, entertainment troupes, archery, clay pigeon shooting, model aircraft and boats. In the main arena there were also many displays ranging from American square line dancing, falconry, road rescue vehicles and many more attractions.

One from the John Ward stable, this 1976 ERF 5MW looked splendid, arriving early to the site and all loaded up which makes these lorries look even better!

Weston Park is situated on the Shropshire/Staffordshire border, about eight miles from the M6 and just off the A5 between Cannock and Telford. In the grounds there is a splendid stately house, which was built in 1671 and has been the home of the Earls of Bradford. Today the house and park is a charitable foundation, the family's collections of porcelains, antique furniture and portraits are available for all to enjoy. The ongoing restoration and maintenance is financed by events such as this one.

The show is indebted to all those companies and their drivers who have joined together to present such an impressive display from the 'World of ERF', not forgetting the organisers of the show, the pERFectionists.

Having arrived nice and early at the site on the Sunday morning, I decided to position

This former fire engine has a low recorded mileage. Converted to a recovery vehicle, it is owned by Philip Nixon of Crewe.

One ERF that I have seen at many occasions is this 1954 flat, NWY 716, owned and driven here by John Gascoigne, Sheffield.

myself alongside the road leading into the park, to capture on camera some of the ERFs as they arrived. Quite a few had arrived the day before, but Chris Smith, the organiser, told me that there would be many more, and he was not wrong. Along came Tony Davies in a very nice example of a 1968 'A' Series flatbed, FMB 208G, in splendid condition, having done a million miles and still going strong. The little flat is in the livery of Ray Grocott and was bought new by his mother - what a present!

It's always a pleasure to see something different and the 1970 eight-wheeler box bodied ERF, owned by Philip Potts, started life as a tanker, new to the CWS in Liverpool, from where it passed onto J W Lees Brewery in Middleton. Bought by the Potts family in 1986, they removed the tank and replaced it with a box body, as they used it to transport their vintage tractors to shows and events. Philip tells me that he's replaced the original Eaton drive axle and fitted Kirkstall axles all round. The lorry has a Gardner 180 engine and has done 80,000 miles - a true work horse.

Seen both on the parade and in the park was a KV cabbed ERF, which has quite a history to tell, from speaking to the present owner Geoff Hinchcliffe. This little flatbed has had a few owners including our friend Eddie Timmins, who explained the background to this unique little lorry, and rekindled fond memories of his days running her. It was new to British Sidac at St Helens with the original registration HDJ 534D, which was later changed to RRT 591 under the ownership of Rob Roy Transport. Eddie tells me that he ran it for ERF doing daily runs down to London and back, quite a journey in those days.

It was then passed to a coal merchant in Liverpool, who ran it for a short while, before changing hands again to Rob Roy Transport, and subsequently now owned by Geoff Hinchcliffe Transport. Geoff plans to keep her and intends to obtain the original registration plate and documents to replace RRT 591. The lorry is in great condition, the only work Geoff has done is a repaint and lettering. Apart from that the rest was mechanically sound, and it looks a gem in the

Plenty of back-up with this recovery truck from Beech's Garage seen on standby.

bright red livery, having now covered a genuine 97,848 miles. Eddie Timmins has even got the original operator's log sheets he filled in and the service records, and has offered to hand them over to Geoff. That's the spirit of these chaps! My thanks to both Geoff and Eddie for their help in providing me with the detailed information for this report.

Perhaps the most interesting part of the day for me was chatting with a chap called Jack Spittles of Wolverhampton. Jack is a dab hand at making models of lorries. I'm not exaggerating when I say that Jack's ability to make these models out of sheet plastic is quite amazing, with excellent attention to detail, and the models are to 1/16 scale. They are finished in the livery of a particular

RF
FORDSHIRE

schuk

aulage company from the past. They ertainly created a lot of interest on the day, et out on the back of GMR 93, a 1950 4LK flatbed. It takes Jack about 12 months make one from scratch and over a lifetime e has made many models. Jack prides imself on cutting no corners when it comes detail. After all, the owners of the actual orries themselves are there to criticise if need e, so it's important to get it right! Jack imself has been a lorry driver for the past 60 ears.

It's always a pleasure to see the many ifferent types of commercial vehicles bearing ie ERF badge, and the 1970 ex- taffordshire fire engine is no exception, as wner Philip Nixon explained. It was onverted by the fire service at their Hanley

This 1950 ERF 44 LK is owned by Plant Transport and sat on the back is Jack Spittles, proudly displaying his 1/16 scale models.

workshops into a recovery truck when it ended its career as a fire engine, it then became surplus to requirements so they disposed of it. It was subsequently bought by a chap who used it for local recovery work until the law changed regarding the way vehicles are recovered from a roadside accident, forcing him to park it up indefinitely. Philip bought it in April of this year, but he has since got his eye on another ERF, a flatbed at present used on a farm for storing hay bales. Incidentally, his recovery truck has a Perkins V8 510cu in (8.3-litre) engine in it, which should pull most commercials if required, and has a mileage of 35,000. Perhaps we'll see Philip around in the lorry he hopes to buy. We wish him the best of luck.

Eddie Wilding from Herefordshire brought

Eddie Wilding driving the 1972 'A' flatbed.

What a splendid sight! This 1976 'M' series ERF and lowloader carries 1923 5hp Burrell road loco 'The Badger' and a living van.

The original ERF, number 63 of 1933, leads the line-up before the parade.

along a 1972 'A' Series 45 LG flat, TJL 880K, with its 5LW Gardner engine purring. New to Sellars & Batty of Peterborough, the vehicle then passed onto a fruit and vegetable merchant in Boston, Lincs. It was bought by the present owner J W Delahay in 1993 and restored for use at shows and rallies. There were a handful of other types of classic commercials, including two examples from Alan Sturge. His 1964 Austin dropside tipper, ACT 639B, is complete with a six-cylinder 5.7-litre engine mounted horizontally to enable the driver to walk through the cab - quite a clever idea from the Austin people!

The Austin led a reasonable life, spending most of its time working on a farm in Lincoln, delivering vegetables to markets. Alan bought it at a sale in Grantham in February 1997 and has fitted a new body together with some minor repairs to the cab. The rest was sound. He then bought a second classic - a 1960 Bedford 300 'J' type dropside tipper 813 LMB. This was new to a quarry owner in Bedford and was used for sand and gravel haulage. Unfortunately, this one was in a poor state and as a result needed major restoration, carried out by Alan over a two-year period.

The Bedford has done 114,000 miles and has a load carrying capacity of 8/9 tons. Both these two classics were sign written by Mr Kellet of Stoke-on-Trent, who has done the lettering on all the fleet for Alan and his father since 1952. ●

Sales S

JOHN KILLINGBECK SALE

Report by David Reed

The destination was Blackburn, the reason for the journey was the auction of the entire fleet of Killingbeck Transport Ltd. As the Lancashire town was approached, winter weather descended and the ground was quickly covered with wet snow. Was this the early spring, I pondered, or just another eccentricity of the British weather. Determined to make the sale on time, I drove on through the veil of snow and eventually arrived at the depot of Killingbeck Transport.

Already there were a number of prospective buyers there, examining the tractor units and trailers, the spares and tools, and warming themselves by the stove which eagerly ate up wood at a steady rate. The tractor units and trailers were lined up outside, being steadily covered by snow, while the spares were laid out in the garages. At 10am auctioneer John Harris mounted his upturned milk crate and the bidding commenced.

First it was the spares that came under the hammer and at times John was having trouble getting the bids going. "Wanna bid?" was a common cry, with miscellaneous items often starting at £10. With the firm having run Atkinsons in recent times, there were a number of Borderer spares, for instance windscreens and rear glass were going for around £50 for four, while Fuller 9509 gearboxes were being sold for around £30. There was brisk bidding for a couple of Fuller Twinsplitter gearboxes, one of which went for £425. As the sale continued the number of people with the auctioneer dwindled

Killingbeck's Atkinson wrecker. It was stored for a while before being registered. Originally a tractor unit, it was lengthened by John in the late 1980s and carries full recovery equipment. Reportedly it goes like a train. Bidding stopped at £2,000 and it was unclear whether it was sold.

The line-up of tractor units ready for the auction.

A similar 1988 ERF E14 6x2 tractor unit, this one went for £1,300.

somewhat, perhaps they were waiting for the main event, the sale of the tractor units outside. It was at this time that I managed to have a word with John Killingbeck himself, the owner of the fleet.

John told me that the company was founded in 1919 by his grandfather, J W Killingbeck, who used steam wagons for haulage. In those early days they were developing the general haulage trade around the Blackburn area. Like other firms, Killingbecks were nationalised, but the firm was re-established almost immediately. A lo of work at this time was done out of Presto docks utilising the ferries from Ireland, usin loose flats and unaccompanied trailers. It this shipping tradition that is highlighted o the doors of the lorries, with the Blue Pete flag proudly displayed.

During the 1980s Killingbecks was stronghold of Atkinson Borderers, with 1 Mk2s operating. John thought that they wer a very rugged vehicle which always held the own in terms of reliability. "They gav

Scene

ɔhn Harris conducts the sale of the
arious lots of spares.

nbeatable service," he said. Most had been ebuilt in their garage with Cummins engines eing replaced with Gardners. At this time round £1,000 would have been spent on a econd hand Borderer, with perhaps double ıat being allocated to its rebuild. This nsured that the vehicle was ready for use, sually starting with a trip down to the ocks. Some cabs remain on site, ıonuments to an earlier age of transport. Iore recently the firm have been carrying eer, steel and general traffic.

After a family involvement of 80 years, ɔhn had decided to retire. "I've got 25 years f holiday owing," he said as we continued ur conversation, "It's been a lifetime of ork." He obviously didn't know what to do s the auction progressed, whether to follow ıe crowd or stand and chat. He did however nd some fascinating photographs showing ıe fleet in the '50s. There were pictures of iuys, Leylands, Bedfords and Scammells efore the Atkinsons were introduced. There were some big transitions in the fleet ver the years," said John. One photograph ıowed a number of double-drive Atkis in ıe yard arranged in a letter 'A'. "The largest election of double-drives in one place," uipped John.

After lunch, with the snow at last elenting, it was time to move outside for the ıction of the tractors and trailers. The latter ection was represented with a wide variety of makes, with a 1985 Broshuis tandem axle 40-60 Trombone trailer selling for £4,000, 1970s Crane Fruehauf tandem axle 40-60 Trombone Trailers going for £2,700, with tri-axles selling for around £5,000. 1970s Municipal tri-axle 45ft trailers could be had for around £1,000 while 1980s York curtainside tandem axle 40ft trailers were around the £700 mark, with a 1982 tri-axle going for £1,200. Two 1983 Crane Fruehauf curtainside 38ft tri-axle units with air suspension went for £1,200 and £1,800 respectively, while 1988 Cartwright 40ft tri-axle curtainsiders with air suspension went for around £2,500.

The ERF tractors in the sale included a number of 1988 E14 6x2 units, which sold for about £1,300, while some 1980s L10 6x2 units were knocked down at around the £800 mark. Two 1989 ERF E14 6x2 units went for £5,250 each, including one with a midlift, while a 1991 Seddon Atkinson E14 6x2 sold at £3,500. Two 1994 Seddon Atkinson L10 Strato 4x2 3.8 metre wheelbase units went for £7,450 and £7,500, while a later 1995 model sold for £9,500.

Two recovery vehicles in the sale caught my eye. The first was a 1986 ERF E14 6x4, registered C142 EBU, with full wrecker equipment and winch which went for £3,000, while an interesting Atkinson (TBR 737R) only reached £2,000 when the bidding stalled. This mighty beast had been originally been a tractor unit and was converted to a wrecker by John in the 1980s, including the lengthening of the wheelbase. It is thought to have been made in 1975 but was stored when new, not receiving its plate until a year or so later. It carries an eight-cylinder Gardner engine coupled to an eight-speed David Brown gearbox and Eaton two-speed axle. I'd heard that it went like a train, but when I put this to John he thought that trains wouldn't be fast enough!

This Atkinson cab was complete with the remains of a dead cat.

After the sale of the vehicles it was back to the rest of the spares and miscellaneous lots. There were a number of Borderer cabs around in various stages of decay which were included in the sale, one even contained the remains of a cat that had passed away some time earlier. Response was mixed. There was interest but no buyers for a couple that were piled onto a trailer. At last, the snow started to melt. "We're lucky," was one remark, "it usually blows a gale from over there."

It was evident from listening to various comments that Killingbeck Transport would be missed. "This man's proper haulage," was one that was overheard, "this is how it should be." It's always sad when a long-established firm such as this passes into history, but time marches on, and *Classic & Vintage Commercials* wishes John every success in the future.

Tractor Prices

1989 ERF E14 6x4 E14.400	G750 KFV	£5,250
1989 ERF E14 6x2 E14 320 w/midlift	G410 CSG	£5,250
1988 ERF E14 6x2 E14 320	F504 FJA	£2,350
1988 ERF E14 6x2 E14 320	F465 TBV	£2,000
1988 ERF E14 6x2 E14 320	E969 XNF	£1,300
1988 ERF E14 6x2 E14 320	E973 XNF	£1,350
1988 ERF E14 6x2 E14 320	E376 WNT	£1,400
1988 ERF E14 6x2 E14 320	E377 WNT	£1,300
1988 ERF E14 6x2 E14 320	E379 KBV	£1,200
1988 ERF E14 6x2 E14 320	E229 WKL	£1,350
1988 ERF E14 6x2 E14 NTE 290	E302 LBF	£1,100
1988 ERF L10 6x2 L10 290	F780 FRJ	£950
1986 ERF E14 6x2 E14 320	D678 MNF	£1,100
1986 ERF E14 6x2 E14 320	D920 RHF	£1,500
1987 ERF E14 6x2 E14 320	D366 NKK	£1,450
1987 ERF L10 6x2 L10 290	E462 TEC	£1,000
1986 ERF E14 6x4 E14 320 wrecker equip + winch	C142 EBU	£3,000
1985 ERF E14 6x4 E14 320 ballast	B839 EET	£1,650
1986 Seddon Atkinson E14 4x2 E14 320	C950 OCU	£900
1986 Seddon Atkinson E14 6x2 E14 290	C492 HDM	£800
1983 ERF L10 6x2 L10 290	A74 LEC	£750
1991 Seddon Atkinson E14 6x2 Strato E14 400	H290 DNF	£3,500
1995 Seddon Atkinson L10 4x2 L10 325 T6 Strato	N876 FCW	£9,500
1994 Seddon Atkinson L10 4x2 L10 325 T6 Strato	M123 LOK	£7,450
1994 Seddon Atkinson L10 4x2 L10 325 T6 Strato	M328 PFV	£7,500
1975 Atkinson Recovery	TBR 737R	stalled at £2,000

STEP BACK IN TIME

Some might see it as a form of punishment, but to dyed-in-the-wool lorry people it's a weekend of sheer, unbridled enjoyment. The North of England Two Day Road Run has become the focus of classic truck owners from all over the UK, and in the eleven years since its inception it has grown into one of the best events of its kind.

By Peter Davies

The run itself is approximately 250 miles, but for many entrants you can add on another 500 or more for getting to and from the event. What makes the run so rewarding is the fact that the whole route takes in roads once frequented by working lorries; and there is no sitting around in country parks and fields. Each day the start and finish are both located at Truckstops where the lorries and the drivers are on familiar ground.

There were no less than 105 entrants at this year's Run and there were some superb machines among them. Each year standards improve and even the roped and sheeted loads are becoming more realistic. Sadly, the continuing increases in DERV prices are making such events more and more expensive for the enthusiast. At normal pump prices it's now £120 or more for a typical fill-up, so in fuel alone you can wave goodbye to a couple of hundred quid. This year I was delighted to be given the opportunity of driving the latest addition to Robert Wells' collection of eight wheelers. Having completed the run in a variety of different 'eightleggers' in the past, it was a pleasure to drive something different and more modern than I had been used to. On previous occasions, I've entered my own Maudslay Meritor and AEC Mammoth Major MkIII and had the privilege of driving Tony Knowles' Youngs Express Scammell Rigid Eight. I've also been grateful to John Richardson of Richardsons Commercials, Oldham for the use of his 1951 Atkinson L1586 and that magical AEC MkIII eight wheel tipper HFX429.

The ex-Hobson ERF belonging to Robert Wells and driven by Richard Nixon.

Robert Wells' latest machine, the 1971 ERF tanker which I had the privilege of driving on the Run.

When Richard Nixon, who maintains and drives many of Robert Wells' old-timers, rang me to say that RND650J, a 1971 ERF Co-op milk tanker, was available I was on cloud nine and needed no persuasion to drive it on its maiden run. It was resplendent, having just emerged from a complete restoration in Colin Pitt's workshops at Otley.

The journey up to Otley from Amersham was enjoyable in itself - riding shotgun on Robert's other ERF eight wheeler XDB235K. This is a remarkable machine which, so far, has survived in working guise. Until a couple of years ago it was still in service with owner driver Don Hobson of Alderley Edge in Cheshire, carrying mainly roofing felt from Manchester to the South East.

To keep pace with modern trucks, Don had replaced the original Gardner 180 with a 6LXCT about ten years ago. The turbocharged Gardner puts out 230bhp and, while the standard David Brown six-speed box was initially retained, Don later replaced that too with a Fuller RTO9509 range change. This, in combination with the Eaton two-speed axle, means that the ERF has 18 forward ratios. The shift pattern, Richard explains, takes some getting used to as it's not only a typical Fuller overdrive 'round the

One of my personal favourites was this ex-BRS Seddon, now in Colin Pitt's collection.

Shap summit was lined with lorries as far as the eye could see.

dry and overcast. No less than six lorries set out from Colin's garage at Leeds Road, Otley. Pete Midgeley's Albion Clydesdale looked resplendent in its original H B Wright colours while Colin's entries included his rare Vulcan 6PF, his Seddon Mk5 which he still regards as his favourite motor and my personal favourite the ex BRS 16:Four-470. This looks just as a lorry would have looked and is free from the indignities of varnished timber and girlie names. As such it gets my vote as one of the nicest motors on the Run.

At 10am it was time to fire up the wagons

houses' affair but is also right to left, the lowest gears being nearest the driver. Robert bought the 28 ton gross ERF after seeing it in Nick Baldwin's 'rags to riches' feature in our September 1998 issue.

The ex-Hobson ERF has an impressive turn of speed, so we made good time on the journey up to Otley on Friday May 7, where we met up with the newly completed Co-op tanker. This is a 1971 6.8GX with a Gardner 6LXB, David Brown 657 'box and an Eaton two-speed axle. Colin and his team at Otley have done a first class job of the restoration. The decision on Robert's part to return the ERF to its distinctive white and blue Co-op livery is to be applauded and the end result is most convincing, although the authenticity is marred somewhat by the addition of a fleet name over the nearside headlight. I'm told that that is what people like to see.

The weather on Saturday was a mixture of heavy showers and sunny spells but it began

Pete Riley boots the British Steel Foden FG four-wheeler up the A6 north of Kendal.

The two ERFs take a break at the Plough Inn on the A6 en route to Shap.

J D Crawford's 1974 Volvo F88 is a first class example of authentic restoration.

and the garage forecourt was soon veiled in a haze of exhaust smoke. The thudding beat of the 180 sounded healthy enough as I waited for the air to build up and then it was chocks away and the convoy headed out along the A65 for Skipton. We had planned to make our first stop at the favourite watering hole of Devils Bridge near Kirkby Lonsdale.

When the North of England Run first started, there was ample room for everyone to park up at this attractive spot and to grab a snack and a cup of tea, but in recent years it has become so congested that there's never anywhere to park. Consequently, we drove in and drove straight out again and headed for the Plough Inn just north of Kendal on the A6. After a swift shandy I opted to sit outside on the stone wall to watch some of the old-timers pass by en route to Shap.

Among them was the superb little British Steel Foden S18 four-wheeler, owned by Peter Davies of Hyde, Stockport and driven by Pete Riley. This is an excellent example of authentic restoration and the 'period' effect was marred only by rally stickers which Pete had obviously forgotten to remove from the windscreen.

Rab Laidlaw's duo - the Reid Transport Octopus and Ergomatic Beaver artic - looked most impressive as they stormed past up the gentle slope past the Plough. Meikle Tennant was at the wheel of the Octopus and trailer, while Adam Moir was hard on his heels in the Beaver five-axle artic STB550C. Both vehicles carry sheeted loads that look absolutely right.

Derek Cooper Transport's ERF heavy haulage outfit was a sight (and sound!) not to be missed as it roared past in fine style, the sound of its throaty Cummins reverberating through the Cumbrian hills. Also in fine form and playing my favourite music was that magnificent Chinese Six Mickey Mouse artic driven by Tony Knowles.

Back in the ERF tanker once more, I headed up the A6 towards Shap summit, where a huge gathering of trucks had assembled, lining both sides of the road for hundreds of yards. I shouldn't think the famous landmark on the A6 was ever so busy, even back in the '50s. The turnout this year was better than ever and the standards of preservation have got better and better. Where else could a lorry enthusiast find such a mouth-watering array of heavy metal in this day and age?

It was a special treat to hear such a variety of two-stroke engines on full song as they ground up the approach to Shap summit - Fodens, Commers and even a 6-71 Detroit, courtesy of Willie Flomer of Coventry. He has restored a rare 1954 GMC 863 bonneted artic which was once used in the experimental department of the MIRA proving ground.

The large truck park at the BP Truckstop on Carlisle's Kingstown Industrial Estate was jam-packed with old-timers on Saturday evening. It was a good opportunity to admire the multitude of different trucks. There were no less than 18 makes represented. Leylands, ERFs, Fodens and AECs were the most common makes in that order, and between them they accounted for 58 entries - that's more than half the total. Atkis and Albions were also well represented.

There were no Maudslays or Sentinels at this year's Run, reflecting the rarity of these makes. However, it was a sign of the times that two Volvo F88s were among this year's entries, and very nice examples they were too, especially the newly restored artic tanker of J D Crawford & Co Ltd of Haltwhistle. This has been restored from a very rough state, having spent its later days as a yard shunter. It's encouraging to find that the restoration of lorries from the '70s is taking place so that future generations will be able to see such machines as well as the older tackle.

Everyone took full advantage of the excellent facilities at the BP Truckstop on the Saturday evening. After a good night's sleep I was ready to face another day. The skies were overcast and threatening, but then, we were in England, so it was no surprise. I climbed aboard the ERF and fired up the Gardner, contributing to the 'Patricroft Fog' that was steadily obliterating the entire truck park.

Saturday had seen no real problems with the lorry, apart from an intermittent power steering fault, a sticking accelerator (which was traced to a missing split pin) and a speedo that was reading 10 mph slow in high axle. I wasn't so lucky now as the alternator decided to go on the blink. Attempts to find the problem failed but as the vehicle had new batteries they would survive the 350-mile return trip to Amersham without being charged.

The second day of the two-day road run is by far the more demanding, taking in the man steep hills en route to Darlington via Hexham. Greenhead Bank certainly tests the old 'uns to their limits, and, owing to the wet conditions, one or two lost traction and ground to an unseemly halt on the steepest section. The ERF took them all in its stride and the next port of call was the regular stopping place at the top of Kiln Pit, where the lorries could cool down and welcome cups of tea and bacon butties were available from the catering van specially laid on.

Suitably refreshed, I was back in the driving seat once more, destination Barton Truckstop near Scotch Corner, where the sun came out and we could all have a last opportunity to admire some of Britain's finest preserved trucks before they set off home in various parts of the UK. For Richard and myself the day was far from over - we still had over five hours driving ahead of us. Some of the drivers had even further to go, like Shaun Swain who had brought his AEC drawbar outfit all the way from Carmarthen in South Wales.

Once again the North of England Two-Day Road Run was a resounding success thanks to organiser Ted Hannon and his son Dominic. I might be tiring but I can think of no better way to spend a weekend. I am most grateful to Robert Wells for providing the ERF eight wheeler which performed so well on its first trip out.

Since completing the run, I've bought S Coupland's splendid video so that I can sit back in the comfort of my own armchair and watch the action all over again. ●

Another motor that deserves full marks is this nicely turned out Bedford A5 from P W Archer.

Ted Hannon, seen here with 2½-year old Fraser, can chalk up another great success with this year's North of England Run which he organised with his son, 20 year old Dominic.

NEWS

CLASSIC and Vintage COMMERCIALS

ERF Sold!!!

By Peter Love

Shock news! As we were going to press it has come to light that the UK's proud, Cheshire-based truck manufacturer ERF has been sold off by Canadian owners Western Star. ERF has been bought by their European competitor MAN (Manschinenfabrik Augsberg-Nurnberg) who have been established since 1915.

Western Star sold ERF for £110 million, having paid just £27.4 million back in 1996. The deal for the Sandbach manufacturer was for £65 million in cash and the balance in assumed debts. In 1998/9 ERF had a turnover of about £200 million with sales of around 3,000 trucks.

The deal will make MAN the biggest truck maker in Europe. Incredibly, it was not that long ago that MAN was under scrutiny for a possible take-over by ERF! One wonders what MAN's thinking is behind the take-over, perhaps this is a way of making MAN a more desirable company to be taken over itself. For many years a number of companies have been interested in purchasing it, but it is said that it is just too small to interest a major, a fact that has certainly changed now. No matter what, this is certainly an upturn in events in the roller-coaster world of truck manufacture, particularly for ERF.

We have been told the companies will stay independent of one another and ERF managing director John Bryant insists that it will be business as usual. Nevertheless it makes an interesting scenario, especially if you think back to what happened after Steyr taken over by MAN. Now that leads us to an interesting connection, as various ERFs use that slab-sided Steyr cab. Let's hope ERF do stay independent and go on for many years to come...

You see them everywhere, all over Europe and increasingly so in the UK. Continuing our thoughts on MAN cabs, this 1974 example has the familiar cab which was introduced by the French Saviem concern in the late 1960s.

MAN is becoming more and more popular in the lighter end of the market, but this 1972 MAN GUL motor home is something else! It was spotted in a car park at Ringwood, Hampshire on September 1 1998.

THE LEAVING OF SANDBACH...

Why have ERF and Foden, both now foreign-owned, deserted their traditional home town of Sandbach? Malcolm Bates tries to find the answer from the archives, and ends up resorting to soap opera...

Welcome to Steam Wago Mews. Down 'Mickey Mouse Drive' is 'Two Stroke Crescent'. The new 'Incomers' are already taking u residence..."Bye, Darling, I'm just popping out in the Shogun to 'The Kleer View Retail Experience' for a few imported items of DIY comestibles, back in five... Oh, whil I'm there, shall I bring us back something from 'The Chinese Six' take-away?..."

It's no good, rather than crib a few ideas from 'Dynasty', I wanted to base my script o something called 'real life'. But it just doesn' sound right. You're not that busy for a second, are you? Good. Maybe I could run some ideas by you?

Today 'The Foden Brand' faces a future more a part of Paccar than as part of Sandbach.

The press launch of the new high roof sleeper cab Foden XL range prompted Malcolm's recent trip to Sandbach.

This may look like the 1930s but it's the '70s, confirmed by the Moss Lane works above the railway line. Alas, already, virtually all the old works below it has gone.

Once upon a time, in an attractive little market town, there lived a powerful family of artisans. They were a proud bunch and believed in doing t'job right. While there wer those who had turned a coin importing foreign products - such as 'horseless carriages from a bloke called Benz - the Foden family were of the view that if you wanted a wagon to work reliably in tough conditions, you needed it made from the ground up by Britisl craftsman.

Being a caring employer, the company eve built a little town for its workers alongside th works and for several decades, 'pride' and 'Elworth Works' were synonymous. Things went well and the products of the Foden mily firm went out to several corners of the obe. Well obviously, that's a figure of speech ecause 'a globe' can't have corners. But nyway...

But there was trouble at t'mill. As so often e case, family politics were to blame. But stead of 'JR Ewing', we find 'Black Annie' nd 'ER' - who get exiled to the Antipodes. s in all good TV Soaps, we'll also need a pin off' series. After the family split and xile, comes a plan to set up a rival firm of rry makers, using some more advanced ethods than those being used by the old rm - 'modern' meaning doing less work ourself and buying in more bits from others, pparently. The idea is the new firm will build e same basic product. In the same town. nly the name is... Well, the name 'E R

Only a few original Foden factory walls survive, as new 'executive homes' spring up like weeds.

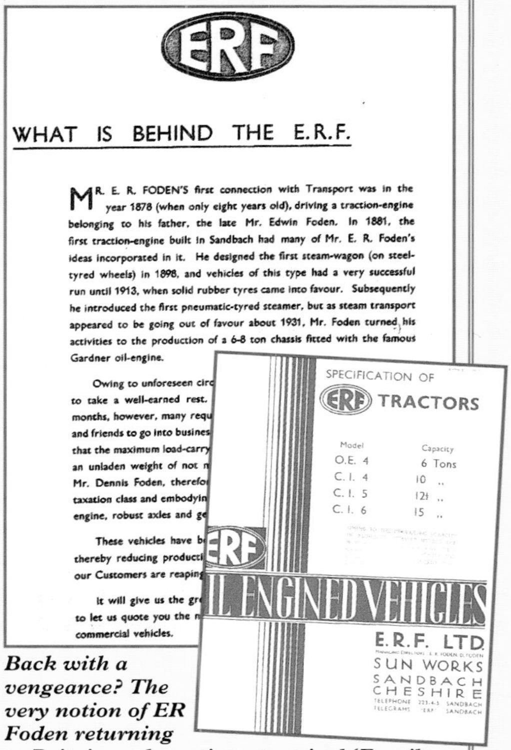

WHAT IS BEHIND THE E.R.F.

MR. E. R. FODEN'S first connection with Transport was in the year 1878 (when only eight years old), driving a traction-engine belonging to his father, the late Mr. Edwin Foden. In 1881, the first traction-engine built in Sandbach had many of Mr. E. R. Foden's ideas incorporated in it. He designed the first steam-wagon (on steel-tyred wheels) in 1898, and vehicles of this type had a very successful run until 1913, when solid rubber tyres came into favour. Subsequently he introduced the first pneumatic-tyred steamer, but as steam transport appeared to be going out of favour about 1931, Mr. Foden turned his activities to the production of a 6-8 ton chassis fitted with the famous Gardner oil-engine.

Owing to unforeseen circ to take a well-earned rest. months, however, many requ and friends to go into busines that the maximum load-carry an unladen weight of not m Mr. Dennis Foden, therefo taxation class and embodyin engine, robust axles and ge

These vehicles have b thereby reducing producti our Customers are reaping

It will give us the gr to let us quote you the n commercial vehicles.

SPECIFICATION OF ERF TRACTORS

Model	Capacity
O.E. 4	6 Tons
C. I. 4	10 ..
C. I. 5	12½ ..
C. I. 6	15 ..

ERF

IL ENGINED VEHICLES

E.R.F. LTD.

SUN WORKS
SANDBACH
CHESHIRE

Back with a vengeance? The very notion of ER Foden returning to Britain and starting up a rival 'Family Firm' in the same town must have been sensational news, but the wording of the announcement and triumphant design of early ERF catalogues (inset) underline how the truth is often stranger than fiction.

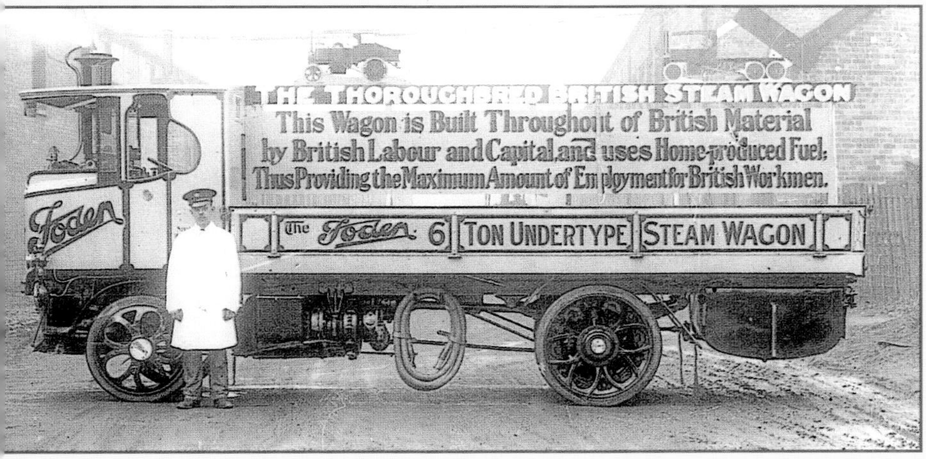

he Bulldog spirit? We've moved a long way from believing 'British is Best' - in fact few the sales points on the banner of this 2-speed E type 6 tonner undertype are true of the rrent product, are they?

Foden & Son' is not totally dissimilar, either, is it? Damn.

Leave that one with me, I'll have to disguise some of that with lots of talk about 'The Great 1930s Depression' or, 'War Clouds gathering over Europe'. Or something.

No, it's no good. Nobody will believe it. It just isn't credible. Two firms making the same product in the same town for over sixty years. Both employing top designers and high quality standards, both competing in exactly the same markets. And both of them staying profitable and British owned - while many of their larger competitors start to crumble due to poor investment, in the face of an imported competition?

Of course to spice up the second series, we could have one - or other - of our two firms threatened by a Foreign take-over themselves.

With the old factory gone, how will future generations of enthusiasts see the significance of the houses once built for Foden workers? The fabric of the place - like the Chip Shop - survives, but the once-proud Co-Op (on the right of the picture) is but an empty hulk

And what future is there for the 'Sun Works' in Sandbach? When it dies, the concept of a truck factory as part of a British townscape goes with it.

Meanwhile at ground level, no longer echoing to the sound of passing 'wagons', the whole area to the left of the road is about to become a housing estate.

That would shock all their loyal followers to their core, surely? But how could it have happened? Let's see. How about one of the firms wins a major truck order from the British Army to the envy of all its competitors? It then has to build a large new facility in order to meet production targets, upsets all its existing commercial customers in the process and then, just when it's up to its neck in overdraft, the same Government puts up interest rates and cuts military spending? Yeah, that should shaft even the best run company.

Then, despite the City of London being the largest financial powerhouse on the planet, there is zero investment interest in British firms making an honest living out of manufacturing things - even those with a strong history and world market potential. So the old firm is hived off to some Americans. Ummm. That might take some explaining. I mean, imagine if the British Hanson empire took over Ewing Oil in 'Dynasty'. There would be a Senate Hearing. But hey, think of the viewing figures...

Trouble is, once you start out on a path of sensationalism, it's hard to keep the ratings up without more cataclysmic happenings, isn't it? So, the next episode sees the other, and until then, strongly independent breakaway company, ERF, also taken over. By Canadians this time. And why not? It'll give me more syndication cheques from Canadian TV...

Big problem. The programme producers and the media types aren't happy. They see the fact that my whole series is set in one little Cheshire town as too restrictive. "Can't one factory be moved, the site sold off for a controversial retail park and all the workers made redundant?", they ask... You know what such people are like - trading on human misery and emotion. I've had enoug of it. I never wanted to be a household nam anyway...

THE TRUTH

Trouble is, the truth is even more amazing than the fiction. Both Foden and ERF - Guvnor's Motors personified - have been subjected to all these trials and tribulations, and more, during their combined histories. But the final chapter is perhaps the hardest to believe yet. I say 'final chapter' with some regret. But then, almost anything is possible these days, I guess. Even a bright future.

As you read this, the bulldozers are tearin through the old Elworth Works site of Fode with a vengeance. Only the odd Victorian factory wall is still standing, as twee little executive homes spring up like weeds. Ther are a few more recent buildings left standin but it's already hard to remember what they once used were for. When I last visited Foden, the whole road outside the works wa awash with activity and the relationship of the worker's houses to the factory buildings was obvious. It was an impressive place. What will future generations of 'Time Team make of it, I wonder? Will we see a future multi-coloured sweater-clad Mick Aston-type dig out a distinctive crown wheel and pinion from under the carport of one of the little boxes, now being built? Will they have just three days to excavate the old Experimental Shop scrap heap? I doubt it.

But just to show you how the pace of progress is accelerating to 'Warp Factor' proportions, the new works across the railway lines in Moss Lane is now also to lose its 'manufacturing' capability. As you read this, the last 'Sandbach' Fodens will have been built and the Moss Lane site turned into a testing and parts storage facility. So where will Foden chassis now be built then?

Here's the twist - at the old Leyland work up the road in Lancashire. This is not quite as daft as it sounds for Paccar, the corporat owner of the 'Foden Brand' (as it's current Managing Director, Dutchman, Aad Goudriaan -ex-DAF - now describes it). After all, the ex-Leyland facility is now also corporately owned by Paccar - after years o long-term doubt, having been treated like a textile industry 'outworker' by DAF.

Interestingly, however, to Paccar, DAF to is just 'a brand' these days. And it seems to matter not a jot that a DAF chassis can just as easily be built in Leyland for export to Holland, and vice versa. And as Foden 'Alphas' now use DAF 85 Series cabs (with an arguably more attractive front panel design), well, no problem, the two 'brands' might as well be built on the same line, Paccar chiefs reason. If this sounds negative then as an enthusiast, I'm not so sure it is - seems that the Leyland works and Foden might well be stronger under Paccar, not weaker.

So what do we have? American owned, Dutch designed chassis now being built in an English factory that once supplied components to the very same fledgling DA

side the Foden erecting shop in the 1920s when the C type steam wagon was king.

Foden started to build internal combustion lorries utilising the marine 5L2 Gardner in 1931 with a Daimler gearbox and a Kirstall axle, chassis no. 14004. Later on the Daimler gearbox was replaced by a Foden unit and a different cab.

ıtfit that grew to take it over and practically ose it down? See what I mean about 'truth' ıd 'fiction'?

IEANWHILE, BACK AT SUN /ORKS...

: the time of my recent visit to Sandbach - inspect the new Foden Alpha 'XL' high of sleeper cab option, as it happens - a run the premises of ERF still presented the ctic and seemingly chaotic mess of what a al lorry factory should look like.

nfinished chassis blasting up the road tween local shoppers and kids going to hool, trade plates flapping in the wind. ew completed units waiting to be shipped f to the dealers and bodybuilders. Trailers ll of components backing up to loading ıys.

Hang on, what's this line of Isuzus doing between the proper lorries? Another zarre twist - ERF now 'assembles' Japanese on-HGV chassis, sells ex-BMC trucks as ade in Turkey, as well as 'narrow track' RF chassis using a cab originally designed Austria for Steyr trucks and a chassis cab nge once produced by Marshalls of ambridge, which had as its basis, a Bedford ıassis configuration. Amazing. And until cently, of course, the same Isuzu chassis ere assembled by Leyland staff on a line at will now build Fodens!

As Foden has announced an ever-tightening of its 'marketing focus' back into the major tractor unit fleet market, ERF is just as busily broadening its marketing portfolio by adding several new types - like going back into fire fighter chassis production for the second time. Only a few years since Peter Foden himself boomed down my phone line that ERF would "Never again" get side-tracked and bogged down with such complex and expensive-to-build vehicles.

At the time he spoke, ERF was itself planning to relaunch itself into the... Well, into the mainstream big fleet tractor unit market with a rationalised range, since you asked. This was later supported by the 'Fuel Duel' campaign aimed at driving the foreign invader back into the sea. Of course by the time that took off, ERF was itself 'foreign' owned - by Western Star.

At that time, it was Foden who was the 'You name it, we'll build it', specialist. Now it's vice versa. Hard to believe and even harder to follow? Well, perhaps. But what will happen next? With the new purpose built ERF factory going operational at Middlewich as we speak, it's clear that all lorry production in the town of Sandbach will come to an end.

But Foden workers who once had their own little 'new town' down the road at Elworth, come off worse - few will transfer to Lancashire and commuting isn't an option. Already, the in-house 'plastics shop' at Foden has gone - as has the last machine shop facility.

But what of the ERF site? The value of 'Sun Works', so close to Sandbach town centre, must dictate it too will be sold off for 'redevelopment'. Remember that phrase about changes happening at Warp Factor speed? Well since I started this article, ERF has passed from being owned by Western Star, to German manufacturer MAN - having paid £65 million for it. This will doubtless bring even more change. So, soon the historian of industrial heritage, or the transport industry archivist will have little evidence of the town's unique past.

There'll be several generations of preserved Foden and ERF lorries of course. The Foden Society and REVS who do so much to keep the names of these makes alive. And shelves full of archive material locked up in the yellowing pages of *Foden News* and *Chassis* magazines. Then, there's my TV Soap script...

But seriously, is that enough? ●

Rival marques? Although photographed when restored in 'works demonstrator' colours, rather than when new, the very first ERF no. 63 of 1933 dropsider had more style than a yard full of Fodens with its Jennings cab. ER was said to have owned and admired a early 1930s Riley Kestrel at the time the prototype cab was designed and copied the window shape.

Sentimental Journey ERF STYLE

Photos: Alec Kermotschuk

ERFs old and new returned to the factory where they were made and formed a cavalcade to the company's new home at Middlewich

Into the 1970s, and this LAV tractor unit leads part of the cavalcade. Christened Billy Boy, STF 404M was new to Northern Ireland Trail (Pandoro) of Fleetwood, Lancs. It has passed through several hands in preservation before reaching the present owners.

ERFs old and new were there to celebrate not just a factory move but the start of, in the words of company chairman and CEO John Bryant: "The second most important stage of ERF's history."

More than 200 ERFs returned to the legendary Sun Works at Sandbach for a weekend of celebrations, mingled with sentiment and some sadness. Celebrations culminated in a cavalcade passing through the town to the state-of-the-art new factory at Middlewich.

The new factory signalled not just change of home, but a new generation of trucks and back-up facilities.

Mr Bryant added that this second stage of the company's history was a natural evolution following the founding of the firm at Sun Works in 1933.

As ERF grew, it became necessary to expand operations and a new facility was set up at Middlewich in 1972. This housed the engineering and aftermarket parts departments.

More recently, cabs have been built at Middlewich.

As Mr Bryant said: "We have for some time had a dream of being able to be back on one site so that all of our employees can be together and, at the same time, give them much more modern and up-to-date facilities to work in."

Looking at the huge gathering of ERF products, including the first and last built at Sandbach, no-one could help but wallow in nostalgia.

The gathering, officially titled the Coming Home - Moving Home event, wa

Seen pulling in to Sun Works is this 1937 C15, supplied new to Alexander Brown of Lanark.

ver ERF mixes with the last example ndbach on cover of commemorative

Seen at Middlewich is DMW 939, an eight-wheeled model 1682, new-to Ushers Brewery of Trowbridge, Wilts, and passing to a showman. It's now owned by S Jones & Sons of Aberystwyth.
Seen behind is a smart C15, new-to Gardner Engines of Patricroft for delivery purposes. It was bought by the present owner's father in 1963 from ERF, who took it in part-exchange for a KV. It's now fully restored.

efinitely a star of the show is this superbly restored KB 4 double-drive tanker, seen in the 1950s Richard Read very.

oking very much ready for action is this model C2TRSVI, built as a two-axle tractor unit and gistered KRS 129V, but since converted into a three-axle ith engine upgraded.

37 model C1561 served for many years with ipperfield's Circus before being rescued for preservation.

jointly owned by ERF Ltd and REVS (the Register of ERF Vehicles Society).

By 11am on the Saturday of the event, lorries were arriving thick and fast, with the classic vehicles forming a wonderful spectacle in the old Sun Works factory yard and the newer vehicles parking up at a nearby school.

A REVS marshall pointed out that 177 vehicles were included in the rally programme, but possibly a further 40 would be arriving as late entries. It was a truly splendid sight, ERFs pouring in one after the other. Peter Foden made an appearance and even posed for a photocall, commenting: "I feel like a page three model!"

By 2.30pm, trucks were lined up according to age and were ready for the parade to the new Middlewich factory.

Amid a roar of engines, the cavalcade left Sun Works and made its way up into the town, with crowds of admirers watching.

Rightfully leading the procession was the first-built, MJ 2711, a model C14, standing for 'compression', 'ignition' and 'four-cylinder'. The chassis number was 63, the age of company boss Edwin Richard Foden at the time, and the lorry was delivered to WF Gilbert of Leighton Buzzard, Bedfordshire, on September 1, 1933.

It was eventually bought back by its manufacturer and restored by apprentices. The vehicle was actually transported on the run by a brand new ECX model, a definite pairing of the first and last ERFs built at Sun Works.

This unusual combination was followed by YD 9325, a 1934 C15 owned by Imerys Minerals of St Austell, Cornwall, but new-to Patten Bros of Ilminster, Somerset. It suffered bomb damage in the Blitz at Plymouth, and a new cab was built. The engine in this vehicle has now covered 1.4 million miles without any major problems.

Making its way along Congleton Road and along the A50, the full extent of the convoy was revealed. Some two miles after crossing the M6 motorway on the A54, the massive new ERF factory could be seen in the distance – a new beginning!

ERF management and local civic dignitaries were there to welcome the drivers and their vehicles into the new complex.

Peter Foden became managing director of ERF in 1960 – at the tender age of 30!

Superbly presented, but still seeing regular use is this 1966 livestock carrier, SLG 504F, the last KV eight-wheeler built. It was later recabbed.

Restoration of this LAG by George Brothers of Larne has just been completed. The truck was new-to John O'Neill of Kells, County Antrim.

C15s were well-represented with T Sage of Chard, Somerset, attending in his superb early three-way tipper, AOD 341, chassis number 414. In the splendid livery of Alexander Brown of Lanark was 1937 example, DMB 658, built in 1937.

Owned by the Passoni family from Winsford, Cheshire, is 1934 OE4 model, GEH 554, built in 1938. New-to Greens Transport of Woore, Cheshire, this vehicle has a special chassis giving it a low-loading height. It is one of the smallest ERFs ever built and has more recently been superbly restored by the present owners.

A good number of 54G models were present, among them 999 NVT, ordered by Beresford Transport of Stoke-on-Trent and exhibited at the 1960 Earls Court Show. It more recently found use with a pig farmer in the Crawley area of Sussex, and was bought by the present owner in 1980 for preservation.

A classic ERF astonishingly still in use on regular trips to South Wales and Scotland for cattle shows and sales is the last KV eight-wheeler to be built. 1966 example SLG 504D is a livestock carrier owned by Graham Simms of Cheshire, and smartly kept.

In 1970, changing Construction and Use regulations enabled ERF to introduce its A-series, a completely new range of tractor units, which had a wider frame to accommodate larger engines. This had an updated cab, with new-style front grille, larger headlamps and twin heaters, all of which gave a more comfortable environment for the driver.

This era was represented at the gathering by several fine examples, including PHY 911M, a Cummins-engined A-series tractor unit built in 1973 and new-to BRS Western.

John Gasgoine entered his A-series tractor unit in the livery of the late Harold Fillingham, OWX 955M.

In 1974, ERF took the world by storm with the B-series model, featuring the SP (steel/plastic) cab. An example represented at the event was prototype B-series tractor unit JRE 389N. Originally in an orange and white livery for ERF and exhibited at the 1974 Earls Court Show, it passed to Beresford Transport for general haulage.

Then, 1978 saw a lowered version of the SP cab fitted on both four- and six-wheel ridged chassis and a handful of tractor units, known as the M-series. L Heal of Bury brought along his fine example, YS 218W, new to Blue Band Motors of Lockerbie.

By 1981, an updated SP cab was introduced, and became the C-series, many of which are still earning their keep today. Again, a large contingent was to be seen at the event.

With the completion of its magnificent factory, new parent company MAN and the introduction of the ECX and ECS trucks in July, ERF can look forward to a healthy future. How fitting that the closu of its old works should have been recorde by such a wonderful collection of classic ERFs.

REVS, ERF and, of course, all the owners who brought their vehicles along, should be congratulated for making the cavalcade such a special event.

For more information on REVs, write (enclosing an SAE) to 6 Graffham Grove, Cheadle, Stoke-on-Trent ST10 1SP. Souvenir programmes from the rally are available at £2 each.

Still giving good service at 20 years old is NEH 25W, a B-series eight-wheeler in regular use to carry bulk animal feed. It can be seen in the North of England, the Midlands and Wales.

Looking very much a traditional working vehicle is this KV 64 from 1957, now in the caring hands of Mark Holford from Southampton.

MARSTON
ERF
184
184
BRICKS
OBM 460

YOU NEED HANDS

by Mike Jeffries

This 1954 Kleer-Vue cabbed ERF has a load of 3000 bricks, meticulously stacked by hand. Imagine having to unload them in lashing rain at the end of a long journey, but it was all in a day's work for the hardy 1950s driver.

Laser proofs of this picture are available from Mike on 01803 882067 or www.transportart.co.uk

Smith's classics

Graham Edge's series on the fantastic restorations in Robin Smith's collection continues with one of the newest trucks we've ever featured, a beautifully presented 1984 ERF C-series, modified for a vital role...

Ready for work as transporter for the slower lorries restored by Robin and friends. The Gardner badge has been retained, although power is now provi by a Cummins 14-litre. The original Gardner 8LXCT unit went the way of many Gardners, finishing up as the pr and joy of a Hong Kong fisherman.

While this ERF is not yet quite old enough to be considered "vintage" if the rolling 20-year rule is applied, it is an important part of Robin Smith's collection. Its purpose is to act as transporter, initially for his Albion CX1, which is rather slow for driving to distant events. This tractive unit has also been considerably modified.

The ERF started life with a day cab and Gardner 8LXCT engine in the ownership of the famous Suttons of St Helens. They double-shifted it for round-the-clock duties hauling chemicals from the ICI plant at Cleveleys. After 10 years and over one million miles, it was disposed of to Walsh and Dearden for scrap.

Robin spotted it from his office window and bought it with the intention of using it to carry his Albion about. It was repainted into Ashworth Products' livery, but when it was finished, it coincided with a busy period for the company.

It was put into service in Ashworth's fleet and after three months, one piston picked up its liner in the engine. This was repaired and a sleeper cab fitted to provide accommodation when needed for the occasional weekend away at a rally. Because a trailer capable of carrying two lorries had by then been acquired, the chassis was also lengthened.

With a trailer, the Albion and ERF tanker on board, the old Gardner was starting to show its high mileage. In particular, it was struggling on its return from the Bournemouth to Bath run in 1998. Robin decided the engine was very tired, and as 8LXCT units are now as rare as snowstorms at the equator, it was replaced with a Cummins 14-litre 350bhp engine. The rear axle ratio for a lower-revving Gardner had also been retained, so this ERF has a fair turn of speed.

Robin quite openly admits that this ERF is non-original in many aspects, but it still plays a vital role in his collection... as well as being an interesting vehicle in itself!

Next issue: A look at a magnificent Albion!

The ERF has been significantly altered from its original specification to be more suitable for its role in Robin Smith's collection. Its chassis has been stretched and a sleeper cab fitted.

eeper cab fitted onto the C-series had been on a lorry engaged oal haulage and the roof lining was very grimy, as depicted by this photo. Much effort was needed to clean it.

A bit of a wiring headache that had to be sorted before the C-series was finished.

Smith's Clas

ERF KV

In his continuing series on Robin Smith's collection of superbly restored classic trucks, Graham Edge looks at an immaculately restored 1961 ERF KV

Robin Smith had always wanted a KV-cabbed ERF – his father had owned one. By a strange quirk of fate, he didn't need to look far to find an example. It was in a yard opposite his own factory, though it had been hidden under a tarpaulin.

Chassis number 9987, this ERF 68GX was completed on June 16, 1961, and supplied as a flat-platform lorry to AR Ibbotson of 164 Worksop Road, Swallownest, South Yorkshire. Registered 6460 WY, the truck was used for steel haulage work from nearby Sheffield.

The ERF remained with Ibbotson until 1977 when it was sold to David Ling, a Yorkshire-based showman who used it for transporting his Thriller Speedway Ride.

After 13 years of showland service, Mr Ling sold the KV to Walsh and Dearden in 1990. It remained in their yard at Darwen, Lancashire, until 1995, when Robin struck a deal with Brian Walsh.

As is the case with some old lorries, at first glance 6460 WY's condition appeared to be quite good, but it turned out to be in a rather poor state.

The chassis was very much past its best, especially around the spring hangers and shackles, and it was badly corroded in places. The considerable repair work was tackled in two stages, the front half completed before the rear was tackled. The wooden framework in the cab was rotten and repairs were needed to the fibreglass panels.

Enquiries meant original owner Mr Ibbotson was tracked down, and he revealed the answer as to why the ERF had non-matching front wings. The truc[k] had been damaged in London, with the result that a new wing had to be fitted. Mr Ibbotson recalled the incident well. The lorry's load of sheet steel ended up all over the road!

The Gardner 6LW engine was also we[ll] worn, and so a strip down and overhaul was carried out. Once this was complete, the unit was repainted. The Kleer-Vue cab was also completely rebuilt, the wooden sections being made by Cheshir[e] based B&C Coachbuilders.

Robin had decided to restore the ERF as an Ashworth's Products tanker, four o[f] his preserved fleet now bearing the livery of this company. This was bought by his father, who had in about 1950 started a

Superbly restored ERF KV. The gleaming tank body gives no idea of just how much work was needed to restore it!

KV as found... unexpectedly in a yard opposite Robin Smith's factory at Darwen, Lancs.

▲▼ Front half of chassis complete before work on rear began.

Cab fibreglass needed much repair. Wings were found to be non-matching. ▲▼

rationing, the pork fat was rendered into lard, and eventually this part of the business outgrew the cannery.

Ashworth's, at nearby Church, was a good business to buy as it made tallow, an animal product used in the manufacture of soap and candles. It's still going strong today.

And so the quest for a suitable tank began. Eventually one was found at Rush Green Motors in Hertfordshire, sitting on a 1961 Scammell. The tank is fabricated from steel, with aluminium sheet covering the insulation.

The 'donor' Scammell had been on fire, meaning the original cork tank lagging was burnt and scarred, and needed replacing.

Following the arduous task of separating the tank from the remains of the Scammell, it was discovered that the sub frame under the tank had suffered. In the end a new one was made, along with some new cradles on which the tank rests.

Finally the huge task was complete and the KV was ready to make its debut on the 1997 Cheshire Run, where the opportunity was taken to photograph it outside ERF's Sun Works at Sandbach.

Very few spectators at this, or other events attended by 6460 WY, will have realised just how much work has gone into this vehicle – particularly the tank, which only an enthusiast with the true dedication of Robin would dare to have even contemplated.

Some of the tank supports had rusted through, and in the end, the entire sub frame was remade.

Before the tank was refurbished, it was placed on the chassis the ERF and measured up for size and fitting.

KV complete. On a cold, still and damp morning, the 1961 KV's exhaust fumes linger, creating a nostalgic, if smelly, inder of halcyon ERF and Gardner days.

Tank on its side showing newly made sub frame and discharge valve.

New cradles were also fabricated for the tank where necessary.

k is made of steel, with aluminium sheet covering the insulation.

Tank sub frame fixed to chassis by U-bolts.

LOG ON *by Mike Jeffries*

The ERF C1.562 six-wheeler had a wheelbase of 16ft 6ins, but its 22ft long body is still not long enough for its load of round timber. It is fitted with 10x20 front tyres to take the extra loading on the front axle, and 36x8s at the rear.

A laser print of this image is available from Mike on 01803 882067. His website is www.transportart.co.uk.

ERF
CN 83 24

SABRINA ST

A unique late 1950s semi-forward-control ERF answered a request from hauliers who wanted to run vehicles with a three-man crew. Only 126 were built, but the lorry became a legend, known as the 'Sabrina' – named after a glamour model of equally ample frontal proportions. Nick Larkin investigates (the lorry!) and profiles the sole restored example.

"Hello Playmates... oh, what a glorious thing to be... a fully-grown-up, busy, busy bee". Hmm... lovable chirpy chappie Mr Askey may have been, but you wouldn't think that millions of men would tune into his late 1950s television show with frenzied anticipation. No, the reason wasn't Arthur's well-known catchphrase or in the hope he'd sing his notorious 'Bee Song' signature tune. It must be said that Arthur was a major star at the time, but it was another show regular who made many red-blooded members of the audience tune in.

Sabrina was, how shall we put it... a sort of Samantha Fox-cum-Jordan of her day: that is a young lady of substantial frontal proportions, though with genuine glamour. In those achingly innocent times, Sabrina's expanses were the subject of as much innuendo as decency allowed. The lady (real name Norma Sykes) was also a pin-up and appeared in gentlemen's magazines, such as the one we've featured here, which included a section where ladies could earn themselves two guineas if their photo was published. A glimpse of a suspender belt was the limit of the sights on offer, by the way.

Talking of magazines, you're probably wondering by now whether this article is appearing in the correct publication. Let me explain – and take you to the less glamorous, but still exciting world of ERF's factory in the late 1950s.

The company was enjoying major success with its KV 'Kleer Vue' models available with a wide range of options. One customer, Manchester-based Wilson's Brewery, made a request which resulted in a legendary ERF model capable of carrying a crew of three. The vehicle was to become instantly recognisable due its semi-forward-control specification and superbly styled bonneted version of the KV cab.

The new model was unveiled in 1957 and would take on various designations; the 54 GSF being five-cylinder, four-wheeler and Gardner-engined, the 'SF' standing for semi-forward control. The six-cylinder six-wheelers were 66 GSFs.

An appropriate, if not too catchy a title, you may say. But someone somewhere – we'll probably never know who, or how – dubbed the vehicle the 'Sabrina,' after Mr Askey's glamorous sidekick. As Carl Johnson, chairman of the Register of ERF Vehicle Clubs, says: "There is so much interest in these vehicles within the register, but we've never been able to come up with the answer as to who coined the 'Sabrina' title."

It could have been a group of drivers, or maybe even a dealer, but the name stuck. There are parallels to Foden's 'Mickey Mouse' cab here, though later on in production, it's understood that ERF used the Sabrina name. It needed to do something; only 126 Sabrinas were built before production ceased in 1963. Many were bought by breweries, the three-seater cab being a godsend for local deliveries, and the six-wheelers were also well suited to heavy haulage, customers including Park gate Iron and Steel of Rotherham. Some Sabrinas were exported, including a vehicl pictured in this article which had an oil cooler fitted for a customer in Turkey.

The reason for the Sabrina's lack of sale has never really been made clear – certainl the six-wheeler had an excellent weight distribution. The strongly built fibreglass cab with aluminium doors was also practical.

YLE

Sabrina shows its front.

Frontal impact of the lovely Sabrina gave lorry its name!

The only Sabrina restored so far – John Thomas's ex-Peter Walker Breweries example.

Trusty Gardner engine needed no work.

There was a price premium to pay over standard models, but you would have thought the superbly styled Sabrina (lorry!) would have made excellent economic sense for labour-intensive local distribution. Perhaps they should have made a dustcart version!

The Sabrina also won critical acclaim, not least when two examples, a 56GSF with a David Brown five-speed synchromesh gearbox, and a 66GSF, with David Brown constant mesh gearbox and Gardner 6LW engine, were made available to the press.

In 1960, slight changes to the styling including the fitting of 'stacked' twin headlights (appropriate on a Sabrina!) and towards the end of production, a small number of Rolls-Royce engined versions were built.

The Sabrina certainly seemed to have much potential, but sadl it never really achieved the market share deserved. Many say the reasons for this a a mystery – just like the source of the 'Sabrina' designation!

THE SURVIVORS

With such a small production, survival rates are obviously low. One example is under restoration with enthusiast Andy Ping, and only one Sabrina, owned by Cambridgeshire haulier John Thomas, h so far been restored. As you'd expect fro John's collection, the standard of work is exemplary.

John's vehicle, 6386 ED, is believed to

The sun highlights Sabrina's wonderful curves.

A wonderful signwriting job by Banner Signs.

y of room for three!

Sabrina had standard KV windscreens.

ave been the second-last example built, eing new to Peter Walker breweries of Varrington. This company's subsequent istory became patently obvious during ie Sabrina's restoration, but more of that a minute. On August 10, 1972, the abrina was sold at auction for £160, assing to Stanley Smith of Selston, Nottingham. He used it on hay and straw aulage.

John spoke to Mr Smith recently, who ad fond memories of the lorry. "He found ie length of the body useful because he ould get a large volume of load on there." Ar Smith used the Sabrina for five to six ears, and then sold it to a Mr Bromley of Nottingham, who employed the vehicle on eneral haulage.

When John bought the Sabrina, it still had Bromley's name on it, despite having been sold to Robinson's Funfair. Then in around 1988/89, it was bought by Arthur Reader, a dealer. "He had it on his farm doing nothing for five or six years," John reveals.

John heard about the Sabrina by word of mouth and went to Tamworth, Staffs, to look at it. He bought the vehicle and decided to have it transported to Shropshire-based restorer extraordinaire, John Boughey.

"We really enjoyed restoring this vehicle," recalls John Boughey. "The most difficult part was the fibreglass cab. Whereas normally you can adapt the panels to the frame, with fibreglass you have to adapt the frame to the panels."

The Sabrina is constructed on a frame of ash and angle iron, the metal on this example having survived extremely well. The cab, in good condition apart from the wings, was stripped and the wood replaced, but most of the iron could be re-used. The wooden roof fillets were also replaced. New aluminium doors were made.

"With fibreglass, there's no alternative but to strip the paint right down," he added. It was during this process that history was revealed: "Above the windscreen was Lord Tetley in his riding gear, drinking a pint!"

The Sabrina was stripped down to the chassis rails, which were in excellent condition, only needing to be shotblasted. After being reassembled with new bolts, the chassis was primed and painted.

The cab was painted in Mason's Carnation red. Ken Thomas – the firm started by John's father, Ken, in the 1940s – adopted this colour when it became a limited company in 1962. This was changed to the present green livery when Fenland Transport of Thorney was taken over in 1968, and its colours adopted throughout the fleet.

John Boughey recalls a painting

Sabrina under restoration photos courtesy John Thomas.

Cab stripped down.

Work well underway.

A complete stripdown to the chassis rails.

Sabrina-cabbed cement mixer needs rescuing.

As found in Tamworth. Sound and in running order.

nightmare: "We put two coats of primer, then three coats of polyurethane topcoat on the cab," he recalls. "But for some reason, the paint reacted on the third coat and we had to start all over again – a week before John wanted the vehicle. We think the reason was a batch of suspect thinner."

Thankfully, the ERF's mechanics proved less of a problem, the Gardner 5LW engine needing no work. The injector pump was overhauled by Les Wall of Dieselec Services, based near John Boughey's Much Wenlock premises. The David Brown five-speed overdrive gearbox and split rear axle were also excellent, and even the brakes only needed a stripdown and clean.

The Sabrina was completely rewired, and an interior retrim was carried out by Telford Intertrim. Thankfully, the ERF's distinctive aluminium mouldings could be reused. John Thomas had a spare 'oval' for the grille left over from an ERF tractor unit restoration.

One item of brightwork did cause problems. John remembers: "The bumper was bent and battered and Keith Wright, one of my chaps here, made a jig up. He kept bolting the bumper on the jig, warming it up and had it sent over to John. He'd send it back with chalk marks saying how it should fit. It must have gone back eight or nine times!"

Finally, the Sabrina was transported to Cambridgeshire, where GB Commercial Bodies of Wisbech built a new body in aluminium with a wooden floor. The back panels and headboard are also in aluminium. The body was repainted at Ken Thomas, Banner Signs of Wisbech doing a fantastic job of the signwriting.

Finally the Sabrina was finished, restoration taking just under two years. "I shall never forget the first time I took the Sabrina to the ERF Gathering at Banbury. People were drawn to it like bees around a honeypot."

Since then, John has driven the Sabrina thousands of miles. "I wouldn't say I have particular favourite in my collection, but I do tend to use the Sabrina on longer runs. The first time we went on the North of England Run, we checked the mpg and it did more than 20 to the gallon. It will cruise at 50mph, too."

The lorry has been on four North of Englands, also the Tyne Tees, Trans Pennine, West of Scotland and many other events, including a visit to Ireland.

"It's run like a sewing machine... touch wood. The only engine problem I've had was a water pump go. It was leaking water inside the engine. Luckily there was no damage."

John has been interested in Sabrinas since they were current. "HE Musgrove and Sons, coal merchants of Stamford, Lincs, had a couple in two-tone green, and they looked superb," he recalls. Ken Thomas's first ERF was an LV with 120 Gardner, JVE 660F. If only that was still around.

John was also attracted to the Sabrina because of its rarity. "It costs a lot to restore these lorries nowadays, and it's good to be able to restore something rare. The Sabrina's so unusual, it's comfortable, quiet, reasonably fast – despite the 5LW's reputation for being slow and economical."

Talking of surviving Sabrinas – or possibly not. There is one bizarre vehicle that John knows does survive. It's a factory built cement mixer with a set-back axle, but is believed to be a conventional 664GX with a Sabrina cab. It was apparently built

Cab definitely a little airy.

As if by magic – Tetley man appears.

Fibreglass had survived well.

so that the water pump for the mixer mechanism, normally protruding from the front of vehicles, could be enclosed.

This strange machine is derelict and its purchase would not be easy to negotiate – but don't be surprised if you see a rather sad cement mixer travelling on a low-loader along the A47 in the direction of Ken Thomas Ltd!

THANKS TO:
REVS chairman Carl Johnson, Peter Davies, John Boughey and John Thomas for their help with this feature. REVS caters for all ERFs.
Details from 6 Graffham Grove, Cheadle, Staffs ST10 1SP, tel: 01538 756956 or fax: 01538 756956.

Now restored Sabrina seen in its Smiths days.

Interior as found.

The Sabrina now under restoration by Andy Ping, seen at Oulton Park in 1993.

(Photo: Carl Johnson Collection)

Twin-headlight configuration was released in 1960. In addition to the radiator, this had an oil cooler as the vehicle for was a customer in Turkey.

Finished Sabrina with 'dummy' load.

Many Sabrinas went to breweries; this one was with Wilsons.

JOHN BOUGHEY'S TIPS FOR RESTORERS

1. Only take on a vehicle you really love.
2. Do what you can afford. Work out how much restoration will cost – and double it!
3. Be careful with painting – take your time.
4. Use a good signwriter. A lacklustre job here can ruin the appearance.

Sabrina seen near Tunbridge Wells as breakdown truck in 1980, and in original form with local firm Flinlock.

Brewery companies were the biggest customers for the Sabrina. This example is seen with Manchester-based Wilson's. Arthur Ingram took the photo, kindly supplied by Carl Johnson.

(Sabrina says: "Ooh, that lorry driver looks nice, though I only mix with professional people nowadays. I'm a nice girl, I am, but I do like a party. Wonder if there's any beer in that barrel on the roof? Any chance of a G&T, dahlings?")

Sabrina Nostalgia

Glamour model Sabrina joins us for a look back at the truck named – or at least nicknamed – after her.*

Well-loaded Sabrina with showman. Sorry, we don't know the location of the photo, part of Carl Johnson's collection.
(Sabrina says: "I hope you weren't referring to me with the expression "Well loaded". There's more stuff on the back of that lorry than in my make-up bag!")

Looking very purposeful is this 66RQSF. This was one of five Rolls-Royce-engined ballast tractors supplied to Parkgate Iron and Steel. (Sabrina says: "Yesss, it does look rather hunky, just like my boyfriend – so you'd better be careful. He likes lorries, too. What is it about you men? At least this has a Rolls-Royce engine. I'd like a Rolls-Royce – a pink one with white rhino-skinned seats. Oh go on then, a Bentley").

lbove: A shot bristling with atmosphere, taken by 'eter Davies in 1964. This 66GSF model was wned by the Park Gate Iron & Steel Co Ltd. It's een at Rotherham.
Sabrina says: "You'd think they'd have cleaned it p a bit. I never appear before the camera unless 'm at my very best, darlings. My audience would xpect nothing less, and that Mr Arthur Askey ould put me over his knee").

★ Sabrina was talking to Nick Larkin. (Errr, be very wary of this. Nick does hear voices, you know – Office Psychiatrist)

'ight:
n extraordinary ehicle believed to still urvive in a derelict tate. This factory-uilt vehicle with set ack axle was a onventional 664X iote the position of ie radiator filler cap) ith a Sabrina cab. was apparently built that the hydraulic ump for the mixer echanism, normally rotruding from the ont of vehicles, could e enclosed.
abrina says: "Now ok, I hope that talk of ings being grafted on n't a dig at me. As for e 'protruding' – well! 's all my own, you iow. All woman I n, dear, though like is cement mixer, I'm one-off. I've got lots impersonators, ough. There's that onroe woman, for a art!").

FINE

More retail nostalgia as we look at Fine Fare's lorries – including a unique restored ERF – and take a quick sneak at J Sainsbury. Nick Larkin reports.

'Where you can be fair to your family, and your purse!' The once well-known slogan of late-lamented supermarket chain, Fine Fare. You might even remember Scottish actor Gordon Jackson preaching it in television commercials during the bleak early 1980s. He used the same tone of voice as when he admonished Rose in Upstairs Downstairs.

You could see the advertising moguls' reasoning for choosing Mr Jackson. Mature man from telly, a Scot, therefore stingy – sorry, someone careful but frugal. It must be true, if he, Gordon, tells us so. Alvin Stardust just wouldn't have had the same impact.

Fine Fare, despite being a highly successful household name, disappeared 15 years or so ago, gulped up by Dee Corporation. Yet the name is being kept alive by a superbly restored 1982 ERF C40, carefully looked after by its owner of 10 years, Anthony Perry, whose dad drove the lorry in service. More on that later.

The Fine Fare brand originated in Canada, but the UK operation was part of Associated British Foods, owned by entrepreneur Garfield Weston. The company expanded rapidly during the 1960s, taking over several well-known grocery chains such as Melias along with regional operations such as Waterworth Bros and WM Cussons.

By the early 1970s, Fine Fare was a highly respected national name, even running some High Street Fine Fare Coffee Bars. The chain managed to cultivate a slightly upmarket image: "Nice stores for nice people," as one commentator said. In less leafy areas, Shoppers' Paradise stores opened with their trademark cheap and frill-free character.

Fine Fare's orange lorries were soon seen everywhere, the company being a major ERF customer. In the 1980s, it was decided to modernise the fleet image and so a changeover to blue began, though this was quashed when the Dee Corporation took over Fine Fare and quickly imposed its Gateway brand.

Fine Fare had a tremendous following in Scotland where it sponsored the Scottish Football League. Thus the name survived

(Photo: Nick Larkin)

It was my dad's truck. Anthony Perry with ERF.

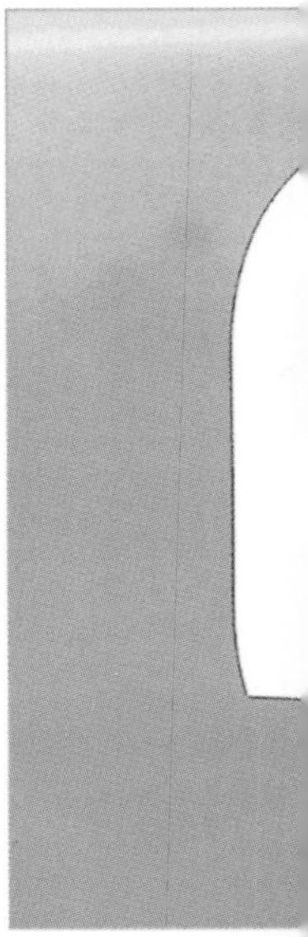

After looking at t... hours, we worke...

(Photo: David Reed)

Restored – and very orange – family heirloom. ERF C-series is now 20 years old.

FARE!

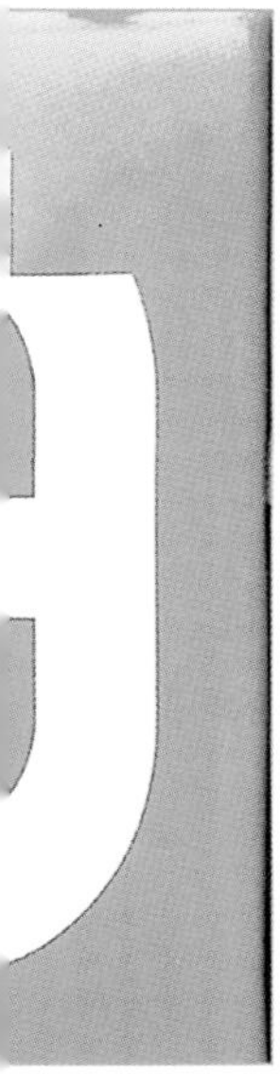

(Photo: Nick Larkin)

...o for several
...d two Fs!

Look at the bargains! Fine Fare newspaper ad, 1980.

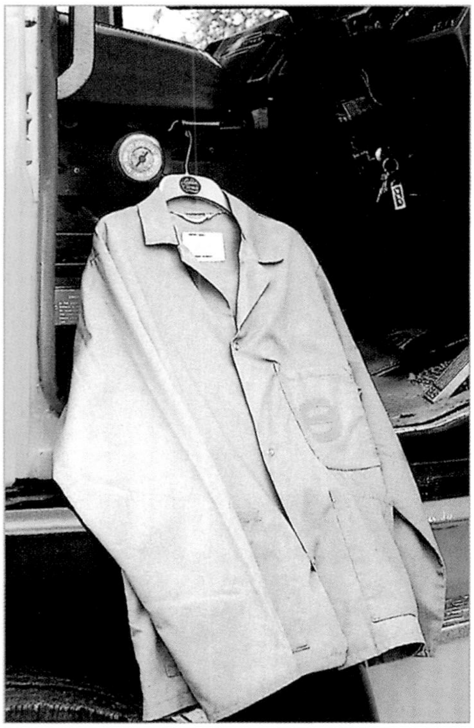

(Photo: Nick Larkin)

Would you wear a Fine Fare jacket for a photoshoot? Anthony reckons it's too small for him. Any excuse!

(Photo: Nick Larkin)

(Photo: Nick Larkin)

New rear wings fitted during restoration.

Cab fittings survived in good condition.

longer there, with lorries that had been painted in the new colours often being transferred to Scotland in exchange for orange ones which would then be 'Gateway-ised'.

A FAMILY CONNECTION

Fine Fare's 'family and purse' slogan certainly rings true as far as the restoration of a unique ERF C-series is concerned. EUR 316Y was driven from new by Malcolm Perry, father of present owner and restorer Anthony. As for the 'purse' – well, change that to 'wallet' because Anthony's carried out a budget restoration of the ERF over the 10 years he's owned it.

Mr Perry senior worked at Fine Fare's Hucknall (Nottingham) depot when he was allocated the bright orange ERF. He'd started his career as a van driver with Burtons of Smithy Row in Nottingham, a regional grocers taken over by Fine Fare in the early 1960s.

Son Anthony, who remembers travelling in the lorry as a lad, said: "My love affair with lorries began when I first went out with my dad in his lorry at the age of five... and every school holiday after that."

The first regular steed was a Cummins V8-engined Ford D Series, OAR 410H. "As a five year old, the noise of the fridge motor used to frighten me, though I'd love to know what happened to this vehicle," he said.

The Ford's replacement was Leyland Buffalo PNG 856R, which had been a demonstrator with Mann Egerton of Norwich. Buffalos were extensively used by Fine Fare at the time – but over the years, Lynx and Marathons were also operated. So were ERFs from KV to LV and E-Series, not forgetting Volvo F7s, Scania 92 and 112s, with rigids including Dodge Commandos and Leyland Clydesdales. As well as Hucknall, Fine Fare had depots at Aylesford, Manchester, Gloucester, Washington, Weedon and East Kilbride.

Finally, both father and son were delighted when ERF EUR 316Y arrived. An ERF C402RT, it was unique among Fine Fare's ERFs in being a C40 model fitted with a Rolls-Royce 290L Eagle Engine, a Fuller Roadranger gearbox and a Rockwell rear axle – the others having the 265L engine and Kirkstall axle. Why this was the case, no one knows.

EUR entered service on store deliveries and was allocated fleet number 9, though this practice was dropped in 1983-84.

When the vehicle was a year old, it was

FINE FARE NOSTALGIA

(Photos: Anthony Perr

(Photo: Ted Short)

Bedford artics were extensively used in the 1960s. This is a TK, 2374 NK, along with step-frame trailer.

Leyland Buffalos were popular with Fine Fare in the 1970s. This pair were shunters at Hucknall depot in 1987.

EUR 318Y was a Seddon Atkinson, seen a Gloucester. It sports the new blue and wh livery for trailers.

Ignore the registration number. This is EUR 316Y starring in the 1984 Goodyear advertisement. If you look closely at the full-sized original, you can see Malcolm Perry's British Rail totem sticker on the rear bulkhead. He used to work on steam locomotives.

(Photo: Nick Larkin)

Malcolm Perry at the wheel of the ERF in its service days.

cleaned and hitched to a brand new trailer for a starring role – in an advertisement for Goodyear G291 tyres, which appeared in numerous truck magazines, the lorry receiving an A-registration for the part. Its 15 minutes of fame over, EUR settled down to a normal life.

Fine Fare had been pioneers of curtainside trailers. The Hucknall depot tended to use step-frame single-axle trailers, progressing to straight box vans with barn doors at 24ft long, then 26ft axles and roller shutter doors, finally 32 ton tandem axle up to 40ft maximum length. Makes used included Crane Fruehauf, York and Craven Tasker. Lorries were generally single shifted, unlike the double and even treble shifting that's the norm today.

In mid-1986, EUR was painted into Gateway green and yellow, then in 1987, the Hucknall operations were taken over by BRS (Midlands) Ltd. There were many redundancies and though the ERF was kept, Malcolm Perry wasn't – he's now a driver with Nottingham City Transport.

EUR 316Y was finally withdrawn in June 1989, then being sold by Cossington Commercials via an auction at Manchester. It then passed to Blurton Transport Services of Stoke-on-Trent. It was uprated from 32 to 38 tonnes, given tipping gear and used on bulk haulage. Due to lack of work, it was later used on local shunting and towing jobs with trade plates.

Meanwhile, all those journeys with his dad, and some work experience in Hucknall, had instilled a determination in Anthony to enter the road transport industry. After his O-levels, he started work as a YTS traffic clerk at Joseph Merritt and Son Ltd, a machinery removals company. He then worked at AET Transport, K&M Hauliers of Hucknall and from 1995, Hays Distribution. In 1998, he won the Transport and Distribution Manager of the Year Award.

It was during a discussion with colleagues that the subject of EUR 316Y came up. Why didn't Anthony try to find it? And so he did – all the way to Blurton

BH 787X, seen at Hucknall, has the blue ery meant to modernise the orange, ich soon disappeared following the Dee rporation takeover.

hoto: Anthony Perry)

Shoppers' Paradise was Fine Fare's no-frills subsidiary, which tended to have shops in less leafy areas.

(Photo: Anthony Perry)

Gateway days, and this DAF 2500 shows off the full livery. Gateway has disappeared after being re-titled Somerfield.

Engine needed little attention during restoration, but cooling fan was renewed.

Bumper stove was enamelled and some re-fibreglassing carried out.

Transport. Armed with a camera, Anthony went to have a look.

"I'd no intention of buying it. I just wanted to have a look for old time's sake," he insists. "While chatting to the owner, it was revealed that the lorry was only used as a shunter, and would be available. I found myself making an offer."

A check by mechanic friend, Neil Utting, confirmed the ERF was in excellent condition, so money changed hands and EUR 316Y was driven back to undercover storage in Nottinghamshire. That was in 1993, and not much was done apart from keeping it in running order, changing the antifreeze and painting the chassis.

Then in 2000, circumstances changed. "I found myself young, free and single, and it seemed the ideal opportunity to crack on and finish the truck. On the horizon was the ERF/REVS Coming Home, Moving Home event, which was an extra incentive."

BJJ trucks near Elford, Tamworth, had a large collection of old/dismantled ERFs in their yard, which produced many useful spares.

"I soon discovered what an all-consuming task restoration can be – especially when you have little practical knowledge of mechanics and are working on your own," he remembers. "That said, the work would never have been done without the priceless help and advice given by many colleagues and friends."

Working to a strict deadline with a full-time job and social life was a tall order, never mind Anthony's other hobby...

A young Anthony with the ERF in October 1985. The location is Beverley, East Yorks.

(Photo: Anthony Perry)

EUR 316Y in a freezing Yarmouth lorry park in the winter of 1982/3.

(Photo: Anthony Perry)

Back on the road but not repainted. ERF in Blurton Transport livery.

playing the tuba in a brass band!

But that did not stop him. First, evidence of accident damage around the nearside step and wing was dealt with by re-fibreglassing. New rear wings and mudflaps were fitted. The fuel injection pump was overhauled and a new secondary filter fitted. The radiator and hoses were also replaced, the front bumper stove enamelled.

Also renewed was the cab tilt pump and pipes, which he found surprisingly to be a relatively straightforward job. New shock absorbers, flasher unit and catwalk were all fitted.

The cab was finally stripped of paint. "Sometimes it seemed I was taking two steps forward and three back," said Anthony. "I was flogging day in, day out, with a small scraper before discovering the benefits of hand drills and Nitromors."

Before he knew it, it was time for an MoT – which the ERF passed first time – and the Banbury Steam Rally. It appeared in a roadworthy but cosmetically unappealing state as the repaint had yet to be carried out. When done, the Fine Fare orange transformed the lorry.

"A phone call to Masons led me to a gentleman in the colour labs who knew the Fine Fare paint codes off the top of his head," says Anthony. Classic Signs and Lines of Mansfield made up the Fine Fare logo and lettering, working from photos.

New keys came easily – local ERF dealer Toptrux in Nottingham was able to get them from the vehicle's chassis number. C40 badges were also supplied.

Now the ERF attracts a lot of attention. "Obviously ERF enthusiasts look at it as a good example of an early 1980s' C-series, but many people find it fascinating in its Fine Fare guise. It was something you saw every day, but has now gone." Anthony is now hoping, storage permitting, to get an original Fine Fare Crane Freuhauf trailer to complete the combination.

Needless to say, Anthony's dad is delighted at his son's achievement. Anthony says: "I've tried to restore it as authentically as possible, though lack of spares and finance has led to one or two non-standard fittings. Now the lorry's looking good, recalling both my childhood and my dad's past. It's a real family heirloom."

EUR 316Y and some sister vehicles just after being painted in Gateway colours.

THANKS

Many thanks to Anthony for supplying many of the photos and much information for this feature. Anthony would like to thank his partner Ann, his dad, friend Neil Utting, East Midlands Commercials,Rob Bradbury, PB Auto Electrics, Toptrux, REVs, BJJ Trucks, Eurotrucks, Masons Paints, Jeyes Engineering supplies and everyone else who helped in one way or another with the restoration.

SAINSBURY ATKI

We're not going to say too much about this former Sainsbury's Atkinson Borderer because we hope to feature its full restoration in a future issue, along with that of Atki owner Clive MacDonald's extraordinary S&D dustcart. But here's a sneak preview.

PJD 817L, a T3046LX model, was first registered in May 1973 as a short wheelbase 30-ton tractor unit, fitted with a Gardner 150LX engine, David Brown six-speed gearbox and Kirkstall 5.75 ratio rear axle.

Supplied to Sainsbury by Atkinson Vehicles (London) Ltd, it had extra equipment fitted to the supermarket's specification – full power steering, Bostram seats and a hydraulic pump to gearbox PTO to run a Bonallack 30ft single axle Concorde fridge trailer.

The Atki was allocated to Buntingford (Herts) depot, servicing stores in the Birmingham and Coventry area – running two shifts totalling 20 hours a day.

Sold to the McKewen Bros, boxing booth showmen of Collumpton, Devon, the fifth wheel was removed, the chassis extended by three feet and a box body fitted with a generator.

After passing through several other owners, the Atki ended up with Clive MacDonald of Surrey in 1996. Clive has removed the box body, returned the chassis to its original length, rebuilt the engine and removed and reframed the cab – a 17-month restoration.

(Photo: Nick Larkin)

Ex-Sainsbury's Atkinson Borderer, as exhibited at this year's Bromley Pageant of Motoring.

CLASSIC and Vintage COMMERCIALS

POSTERS OFFER

Magnificent original commercial vehicle '50s, '60s and '70s colour ads at large A3 size (16"x11") encapsulated in clear plastic for complete permanent protection.

Twenty Two famous commercial vehicle colour ads, to transport you back to the golden age of classic commercials. Printed on high quality art paper encapsulated in a plastic laminate to give them a permanent beauty. Available to readers of *Classic & Vintage Commercials* magazine. PRICE £6.95 each. (All prices plus £1 P&P per print). Please allow 21 days delivery.

Please send me posters as ticked below: *For overseas orders please add 10%.*

1❑ 2❑ 3❑ 4❑ 5❑ 6❑ 7❑ 8❑ 9❑ 10❑ 11❑ 12❑ 13❑
14❑ 15❑ 16❑ 17❑ 18❑ 19❑ 20❑ 21❑ 22❑

ENCAPSULATED POSTERS £6.95 + £1 P&P each

❑ **Complete set of** (22) **£132** + £4p&p

❑ **Half Set** (11) **£71.50** £2 P&P

I enclose my cheque for £______ **(Payable to Kelsey Publishing Ltd)**

(Or) I wish to pay by Access ❑/Visa ❑/Mastercard ❑/Switch ❑

No______________________ Expiry date__________

Issue No. (Switch)__________

Signature ______________________

Classic & Vintage Commercials Posters Offer, Kelsey Publishing Ltd, Cudham Tithe Barn Berry's Hill Cudham, Kent TN16 3AG.
Tel: 01959 541444 Fax: 01959 541400. Mon-Fri 9-5.30.

AMERICAN EXPRESS SWITCH MasterCard VISA

Name ______________________

PLEASE PRINT CAREFULLY

Address ______________________

Post code ______________________

PHOTOCOPIES OF COUPON CAN BE ACCEPTED

ERF BK

BUILT IN BRITAIN ERF SUPPLEMENT

ERF

For seven decades ERF lorries represented the best of British manufacturing. Peter Davies profiles the company and its products.

With the news earlier this year that ERF production was to cease in the UK the time has come to accept that Britain's once thriving truck industry has all but vanished. Of all our manufacturers, ERF shone as a beacon of British tradition for almost seven decades and up to the time that the company was sold to Western Star, just six years ago, it could proudly claim to be the only surviving British-owned lorry manufacturer.

While there is little room for sentimentality in business, the news of ERF's demise this year still came as a bitter pill to the haulage world and to enthusiasts in particular. While ERF-badged vehicles will still be marketed in this country for the time being, they will be little more than badge-engineered MANs. Here we can look back at the history of one of Britain's best known heavy vehicle builders.

Edwin Richard Foden was the youngest son of Edwin Foden who founded the famous Foden steam wagon company whose history dated back to 1856. By the early 1930s Fodens were struggling to sell steamers and 'ER', as he was called, was firmly convinced that the company should build diesels instead. His brother, William, had retired in 1924 and emigrated to Australia. A disagreement over policy led ER to resign and he and his son, Dennis, set up their own company called E R Foden just a short way away from Fodens' factory.

Among the other ex-Foden men to play an important role in the new company were Ernest Sherratt, the chief designer, and George Faulkner, who was works manager. Just after Easter 1933 the four men met in the humble setting of the conservatory at Hilary House, Sandbach, the residence of ER's eldest daughter Flo. Their plan was to shake off the old ways of working that they had become tired of at Fodens and start a fresh approach to truck building.

Partly as a chosen policy and partly because of the daunting cost of in-house component manufacture, they decided to buy in all major units and become assemblers rather than full-scale manufacturers. In doing so they enjoyed greater flexibility which enabled them to respond more readily to specific customers' needs. It was an approach which was later to be adopted by many other major manufacturers especially in recent times.

Production began in 1933 with a modern 6/7-ton payload forward control four wheeler powered by a Gardner 4LW engine. The lorries sold well and very soon the 'E R Foden' badge was altered to 'ERF'. During the six years up to the outbreak of World War 2 the range expanded to include 3 and 4 axle rigids and 2-axle tractor units. ERF was also the first company to build a twin-steer six wheeler.

This was in response to a Scottish operator's request to enable a large capacity Luton van to run without the risk of overloading the front axle. Such twin-steers became a significant type in the UK over the next three decades or so and even today have a role to play. They have other hidden benefits such as greater safety in the event of a front wheel blow out and can offer a worthwhile payload increase over a four wheeler, without a significant increase in unladen weight.

ERF was one of the few UK truck manufacturers to be allowed to continue production during the war and some wartime models were fitted with AEC engines when Gardners were in short supply. After the war, in 1948, a completely new range was introduced under the general banner of the 'V' range, masterminded by chief engineer Ernest Sherratt. ERF went from strength to strength as well as developing an extensive export business.

Domestic vehicles had Jennings coachbuilt cabs, Jennings being an old established coachbuilder with whom ERF shared their premises at Sandbach. Later, in 1963, ERF absorbed the company as their in-house cab builder. For export a steel cab was introduced in 1951, produced by the Willenhall Motor Radiator Co. This was also available on the home market but after a couple of years a completely new and strikingly modern cab with panoramic windscreen and oval radiator grille was introduced as the 'KV' cab. A semi-forward control version, nick-named the 'Sabrina', was added in 1958.

The appearance of the KV at

Right: The very first ERF lorry, Chassis No.63, was this C.I.4 with Gardner 4LW engine. New to W F Gilbert of Leighton Buzzard in August 1933. it was restored by ERF apprentices in 1972 and signwritten in ERF's own name. Chassis number 63 was chosen as it was Edwin Richard Foden's age when the lorry was built.

the Scottish Show in November 1953 added to ERF's prestige and the new models featured many improvements, especially in the braking system. ERF also became involved in building specialised export models for heavy duty operation and options of Cummins and Rolls Royce diesels began to be offered during the late '50s. South Africa became a particularly strong export market for ERF.

The next major launch was the LV-cabbed range of 1962. It was with this range that ERF became the first UK manufacturer to offer fail-safe spring brakes. With increasing demand for articulated vehicles in the UK, ERF began offering a complete matched artic during the '60s. Export demand continued to be healthy and an assembly plant was set up in South Africa in 1965 to build vehicles tailored to local requirements.

In 1969 Ernest Sherratt retired as director of engineering and his replacement, an ex-Chrysler man, launched the 'A' series which was based on a new design philosophy more akin to mass production ideas, especially in the design of frame and suspension. Engines included the Cummins NH220 and Gardner 8LXB 240. The 'A' series formed the basis for a 38-ton 'European' tractor with a steel Motor Panels tilt cab, the 7MW, which ERF aimed specifically at the European market. Soon after, ERF's first true tilt cab range, the 'B' series, was announced. This featured the 'SP' (Steel/Plastic) cab produced by a revolutionary new moulding technique.

Previously ERF's glass fibre cabs were produced using hand laid material in an open mould – a painstakingly slow and not too accurate process. The SP cab was built from hot pressed sheet-moulding compound which gave a better finish and greater accuracy to the plastic panelling. The panelling was bolted around a sturdy steel framework supplied by Motor Panels. It was a successful design of cab which was highly regarded by drivers and the production techniques were still used in the very latest SP5 cabs fitted to the EC range.

Continuous improvements have been introduced to the ERF range since the mid '70s resulting in the 'C' series and later the 'E' series, all using developments of the 'SP' cab structure. The last major launch was in 1993 with the 'EC' range which soon become a leading seller in the UK. A top of the range heavy tractor unit for long haul operations appeared as the 'Olympic'. Since the MAN take-over the EC cab underwent an extensive face-lift to become the ECX. At the same time an optional steel cab based on MAN's design was made available.

In 1996 ERF, Britain's last independent truck builder, was bought out by Western Star of Canada. After the take-over certain ERF-based models were marketed as Western Stars, like the 'Commander' range which sold well in Australia and New Zealand. The mechanical content of the 'Commander' series was predominantly American sourced. ERF continued to offer a comprehensive truck range in the UK throughout the '90s and also began marketing the Turkish–built BMC 'Professional' as the ERF 'EP'. For a time ERF also assembled Isuzu medium weight trucks.

In August 1999 it was announced that ERF production was to be transferred from Sandbach to nearby Middlewich. After 66 years the famous Sun Works was to be demolished. The following February news broke that Western Star had sold ERF to the German manufacturer MAN Nutzfahrzeuge AG.

ERF enthusiasts from REVS (Register of ERF Vehicles Society) and elsewhere gathered at Sandbach in August 2000 to pay homage and to witness a huge cavalcade of ERFs vintage and modern which was staged for the occasion. Though MAN had given assurances that the future of ERF's British manufacturing operations was secure, it later emerged that the plant was no longer commercially viable.

Thus came the sad announcement that, as from July, production of MAN-based ERFs would be transferred to the company's European plants. This marked the end of ERF's 69-year history as one of Britain's best known lorry manufacturers. At least, as enthusiasts, we can enjoy the history of a marque which was once advertised as 'The World's Best Oil Engined Lorry'.

Chassis No.63, the first ERF, just prior to delivery in 1933 with a group of ERF personnel posing proudly in front of their handiwork.

Cover picture: A 1983 painting by Peter Davies, featuring a V-fronted 1954 ERF 6.8 of JG Fielder (Haulage) Ltd.. depicted in 1957.

ERF Ltd

A very early C.I.4, actually Chassis No.65, painted up as a works demonstrator. It was later sold to a Hertfordshire haulier, A Saunders & Son of Harpenden who continued using ERFs well into the 'fifties. Note the full name 'E R Foden & Son' on the radiator. This was later abbreviated to 'ERF' partly because Fodens Ltd objected to the Foden name on what was a rival make. Note also the emphasis on 'diesel' – the main reason Edwin Richard Foden and his son Dennis set up ERF was frustration over the family's persistence with steam wagon production.

Twin-steer six wheelers were a new development in 1936 and ERF pioneered the typ with this large capacity C.I.561S Luton van for I & R Duncan of Edinburgh. Designer Ernest Sherratt explained that it was basically the fron half of an eight wheeler married to the back hal of a four. Very soon many other British lorry manufacturers began offering them and they became popularly known as 'Chinese Sixes' or 'Four in Hand' six wheelers. They took a nominal 10-ton payload.

In 1936, just three years into production, ERF adopted its third style of name badge. This time a vertical arrangement positioned on the top corner of the radiator grille. It can be seen on this impressive rigid six, probably a C.I.562. The '2' suffix indicated double-drive. The coachbuilt cabs were built by Jennings on the same site as ERF's Sun Works.

ERF Ltd

ERF joined the growing number of eight wheeler manufacturers towards the end of 1934. AEC and Armstrong Saurer had already entered the market in that year. Early ERF rigid eights, designated C.I.682, were on a 15ft wheelbase – equivalent to that of a short wheelbase tipper by 1950s standards. This impressive example with a Jennings 20ft double-dropside body was the third machine built and went to W T Parrott of Luton.

Registered on 1st January 1940 this splendid C.I.682 was from the well-known fleet of W H Bowker of Blackburn.

Author

Though photographed in May 1965 this short wheelbase ballast tractor of the Union Cartage Company was actually registered new in 1938. It is seen passing Canning Town railway station in East London. Such machines were used to haul meat containers on four wheel drawbar trailers and were confined to the dock areas and wholesale markets.

Short wheelbase tippers with a choice of 4- or 5-cylinder Gardners, designated C.I.4 or C.I.5, formed a substantial proportion of ERF production in the mid 'thirties. This is a particularly fine example of a C.I.5 with a three-way tipping body. Note that the ERF badge is now in the form of an oval.

'his work-worn veteran from the Sankey of Wellington fleet was hotographed in 1964 with a load of 20in lorry wheel centres parked up in 'oscombe Road, Dunstable where the Bedford and Commer truck ıctories were located. It is a 1950 model and features the stylish Jennings V" cab introduced in 1948.

Author

A particularly fine example of a V-cabbed artic photographed along Chaloner Street, Liverpool in 1964. Dating from 1952, NKA168, is in the bright yellow colours of White Tomkins and Courage Ltd, manufacturers of Clarendo feeds, as clearly stated on the rather stylish bulk trailer.

well laden ERF '6.6' model of Reliance Motors (Bolton) Ltd parked up ılong Liverpool's dock road in the mid 'sixties. The lorry dates from 1952 nd is seen coupled to an 8-ton drawbar trailer.

Author

The Cleveland Products Co of Middlesbrough was the operator of this well laden 5.6TS photographed in Liverpool Docks in 1966. Dating from 1953 it is a particularly nice example, typifying haulage vehicles of the 'fifties era. In the background is the Mersey Docks & Harbour Board canteen.

'he Abbey & Holyrood Breweries of Edinburgh, better known by their 'Youngers' brand name, operated a fleet of fine ERFs for many ears. None finer perhaps than this 1955 V-cabbed tanker. V-cabbed models were giving way to the new oval grilled KV models at the ime this magnificent machine took to the road.

Author

V-cabbed 6.8 models were very popular on bulk tanker work. Here is a nicely turned out 1951 model in the mid-blue livery of Kraft Foods, Edible Oils and Fats Division. It is seen passing the Dolphin Hotel in Liverpool's dockland.

(Author's collection)

Saddler's Transport of Edinburgh were the owners of this early KV-cabbed 5.6TSG of 1955 vintage. This is the first pattern of KV cab before an increased cooling area was provided by additional narrow grilles above and below the oval. All of Saddler's grey and maroon vehicles carried fleet names. This one is 'Kestrel'.

Author

ERF's most memorable cab was the KV, introduced in 1954 to replace the V. For its time its styling was ground-breaking with its unusual oval grille and large curved windscreens. Easingwood's of Blaby were well known ERF users and their tipper fleet always looked smart in its blue and red livery with gold writing.

There were many variations of the KV cab. This superbly turned out eight wheel tanker from the renowned Smith & Robinson fleet features a 2KV12 with set back front axles. It is powered by a Gardner 6LX '150' and dates from 1962. A number of S & R's ERFs with this cab design were powered by Rolls Royce B.81 8-cylinder petrol engines and were designated 8.8R models.

Author

1962 saw the introduction of the impressive new LV cab to replace the KV. The large single piece windscreen was quite sensational at the time and presented the manufacturers Triplex with quite a challenge. This is the Jennings built version of the cab, the LV(J) which, on balance, was probably the nicest looking version. Boalloy pioneered the design for ERF. This fine example of a 6.8GX was new to old established Manchester hauliers W F Sherran in their post nationalisation form of W F Sherran (1953) Ltd of Stalybridge.

From the same era as the KV and sporting an oval grille, this half cab dumper based on a 5.4G chassis has a Jennings built steel cab called the FV.

(Author's collection)

This mid-1959 registered 6.4GX was photographed on the A46 passing through Newark-on-Trent in November 1965. It is in the red and white colours of the Mansfield Standard Sand Co, suppliers of foundry moulding sand. It is coupled to a short twin-oscillating-axle tipping trailer typical of the period when artics were limited to 24 tons gcw.

'he most unusual version of the KV family ›as the striking semi-forward control .4GSF type. They were aimed primarily at ›rewery fleets which required additional ›rew accommodation. This ex-Walkers of Varrington vehicle is in the maroon livery ʼf its second owners, Parry's of Lymm.

(Author's collection)

The annual Lorry Driver of the Year Finals was always a good occasion to see lorries all spruced up. Here is a fine example of one of Showerings large ERF fleet taken at Bramcote in 1973. The 6.4GX is Fleet No.60.

(Author's collection)

efferson Bros of Failsworth, Manchester operated this impressive 6.8GX flat with a 20ft drawbar trailer. ERFs were never as popular for 'rawbar work as their Leyland and AEC rivals so this is quite a rare outfit. Note the obligatory piece of sheeting covering the driver's ›ide of the grille in an effort to get a bit more warmth out of the notoriously slow heating system. KVs were among the first British lorries › have heaters as standard.

TRUNK -BY-
BUCKLEY'S
Marne
Marne
ERF
EM 3349

UP THE ARCHWAY

Evoking nostalgic memories of London in the late 'fifties, this scene depicts the capital's famous Archway Road with Highgate Archway as a backdrop to one of John Buckley's old ERF artics. In those days this long steady climb was busy with heavily laden lorries setting out north in the early evening. The atmosphere was filled with the roar of oil engines and heavy with exhaust smoke. Note Buckley's novel 'elephant's head' nameboard signifying trunk services. Most of Buckley's fleet were named after naval fighting ships – this early post war ERF was named 'Marne'.

Author

The early LVJ cab is also featured on this nicely signwritten 6.4GX from the fleet of Pollock (Scotrans) Ltd. Pollock's inimitable livery sets the cab off beautifully. The vehicle bears the fleet name 'Thunderbird' and is seen laden with a steel coil at Vauxhall Motors H block lorry park in August 1970.

Author

Author

MW-cabbed tractors were perhaps more at home in New Zealand and Australia. This much celebrated machine named 'The Duchess' is known to enthusiasts at home and abroad. The long serving stalwart is owned by Ted Ashforth of Auckland and has only recently been retired. In fact it is currently advertised for sale in the hope that an enthusiast will save her from the scrap man. This picture was taken at Auckland's container base in February 1998.

Often erroneously referred to as 'A series' eight wheelers, probably on account of their 7LV cabs, the eight wheel rigids actually retained the pre-A series chassis technology. They were designated 6.8GXB18O or 6.8CU220 depending on engine choice and were plated for 30 tons gvw.

Author

Successor to the LV-cabbed range of the 'sixties era was the newly engineered A series which took an updated version of the 5LV cab designated the 7LV. Announced the A series featured radically different chassis engineering with supple, outrigged overslung rear springs grouped air tanks and a choice of 8LXB '240' Gardner, Cummins NH220 or Rolls Eagle 220. Design weights were up to 36 tons gcw. This nicely liveried 1975 6.4CU220 with York tandem Freightmaster was from the old established Scott's of Oldham fleet.

nd the same period as the fibreglass cabbed A series was being built offered a Motor Panels steel cab option mainly for overseas. Designated MW it was occasionally seen on British registered vehicles like this fine ng 1974 outfit from the fleet of Turner Bros of Earl Sterndale, Buxton.

Author

Many B series units had centre lift axles fitted to up-rate them to 38 tonnes in the mid '80s. This 1980 registered veteran had covered some miles in its day and was still hard at work in the Walker Trowbridge fleet in March 1992.

Author

When ERF wanted to compete in the solo 16-ton market it brought in the M series which had many unusual design features. It proved to be too heavy and expensive and did not sell as well as hoped. This rare example in short wheelbase tipper form has the standard de-rated Gardner 6LXB developing 160 bhp. A few similar machines were built with Dorman 8F V8 diesels. These early M series, forerunners of the lighter Perkins M16, had remote gearboxes and were only built from 1978 to 1983.

e: A little out of the ary perhaps but an ERF all ıme. This shot is included nind us that ERF were ficant fire engine suppliers een 1966 and the mid-70s Cheshire Fire Engineering ne established at Winsford. HCB Angus-bodied escape n 85ft Simon Snorkel aerial ulic platform and is red by a Rolls Royce B.81 inder petrol engine. When graphed it was owned by Mullard Radio Astronomy rvatory in Cambridge e it was used to clean the ive receiver dishes.

Author

: The first completely new l to hit the headlines since V was the 1974 tilt cab B s. The first off were rigid wheelers. The type is plified to perfection in this rbly liveried machine from mous Brian Harris fleet. It J951T 'Widecombe ıer' and is seen on the A74 Hamilton, Glasgow on 28 mber 1985.

Author

A '180' Gardner 6LXB powers this rare 6.6TSGXB Chinese Six unit coupled to a tandem axle tank trailer in the Sutton of St Helens fleet. For sheer good looks, in lorry terms, there are few vehicles to touch these sensational 5-axle outfits which only appeared for a relatively short period. With its set forward 5LV cab featuring Sutton's unique fleet number 'knob', it was one of nine similar outfits operated by the company.

Author

Most readers will remember the E series with the SP4 cab. Indeed there are some still in operation. Shown here is a mid-'80s classic 4-axle outfit hauled by an E10 sleeper from the famous Onward Transport Group, based in South Yorkshire. It is powered by a Cummins LTA10-290. The official designation of the model, in this case, is a 38C2TRSS1.

Author

It might seem rather ironic, in view of MAN's recent decision to close down UK ERF production, that ERF began doing business with MAN's Steyr Division back in 1988 when it was independent. ERF introduced the all-steel Austrian designed cab on its ES6 and ES8 models. This is a good example of the first style of ES cab on a 16-ton gross tipper. MAN bought out Steyr in 1991 and subsequently bought ERF in 2000.

Author

One of the UK's best known ERF fleets is th Read's are also ERF agents. This nicely liv its air management kit and Perkins (Rolls

Author

No one can rival the Scots when it comes to wheel curtainsider testifies. It is in the exce a direct development of the B series and to

Author

When this E6 twin steer was new in 1987 it was taken, it had been re-engined with a 'I by its second owners, Richardson Roadfrei from Eclipse Motor Transport of Hull.

ead of Longhope in the Forest of Dean. is an excellent example of the breed with X335.

Author

The Steyr cab was face-lifted in 1992 and given a cleaner appearance and a black grille. This highly unusual rigid-eight mixer is not part of ERF's standard model range. The Steyr cab normally only went on four and six wheelers. R111TOP, was a special order and was built on a short wheelbase for 28 tons gross. It was based at Weston Super Mare when this shot was taken in April 1999. It has the Cummins C series 8.3 litre diesel.

lorries as this impressive C series eight cBeans from Kirknewton. The C series was ersion of the SP (Steel/Plastic) cab.

Author

The Cummins M11 engine became a popular choice in the EC range when it replaced the L10 in 1996. This smart rigid eight wheeled EC11 is part of the Interfuse fleet based in Syston, Leicester and is engaged on the distribution of the company's building blocks. It has a centre mounted Atlas self-loading crane.

TA Cummins. By 1998, when this shot ersion and fitted with a curtain-sider body . Richardson's bought the vehicle as a flat

Author

For six wheel rigid versions of the EC like this example in the livery of Booles of Stockport, the C series 6CTAA-240 8.3 litre provides adequate power with 240 bhp gross. The lorry is plated at 24 tons.

Author

Author

Above and Below: After Western Star sold ERF off to MAN in February 2000, ERFs began appearing with the MAN steel cab, designated ECS models. Here are two examples with the distinctive style of radiator grille which disguises the cab's MAN origins. MAN's mechanical units were also being introduced, including rear axles on certain models.

Author

ERF fought without success to hold on to its traditional SP cab, which was highly regarded by most UK customers but it was decided in 2002 to cease production of ERF-sourced models in favour of MAN designs. The ECX shown here is the last design of ERF's long running SP cabs, first launched with the B series in 1974.

Author

To achieve the maximum legal weight various axle combinations have been tried but few can compare with this stunning twin-steer outfit which was the creation of John Killingbeck who was well known for running 'home grown' machines until his company's closure in 1999. Back in 1988 when this outfit was in operation the Blackburn-based fleet was engaged largely on the haulage of steel and beer. This is very much a one-off special and ERF never offered a twin-steer tractor of this type.

'epresentative of ERF's EC range ıunched in 1993 is this 1999 EC11 ·om the fleet of Castle Cement based t Ketton near Stamford. It is hauling Feldbinder bulk cement tanker with design gcw of 44 tonnes. The M11 ngine output ranges from 303 to 399 hp according to model.

Author

With the cessation of UK ERF production MAN TGA types are to be marketed as ERFs with just a minor front end disguise and the availability of Cummins power. This is the ECT (the T comes from TGA) as exhibited at the 2002 Commercial Vehicle Show at Birmingham's NEC.

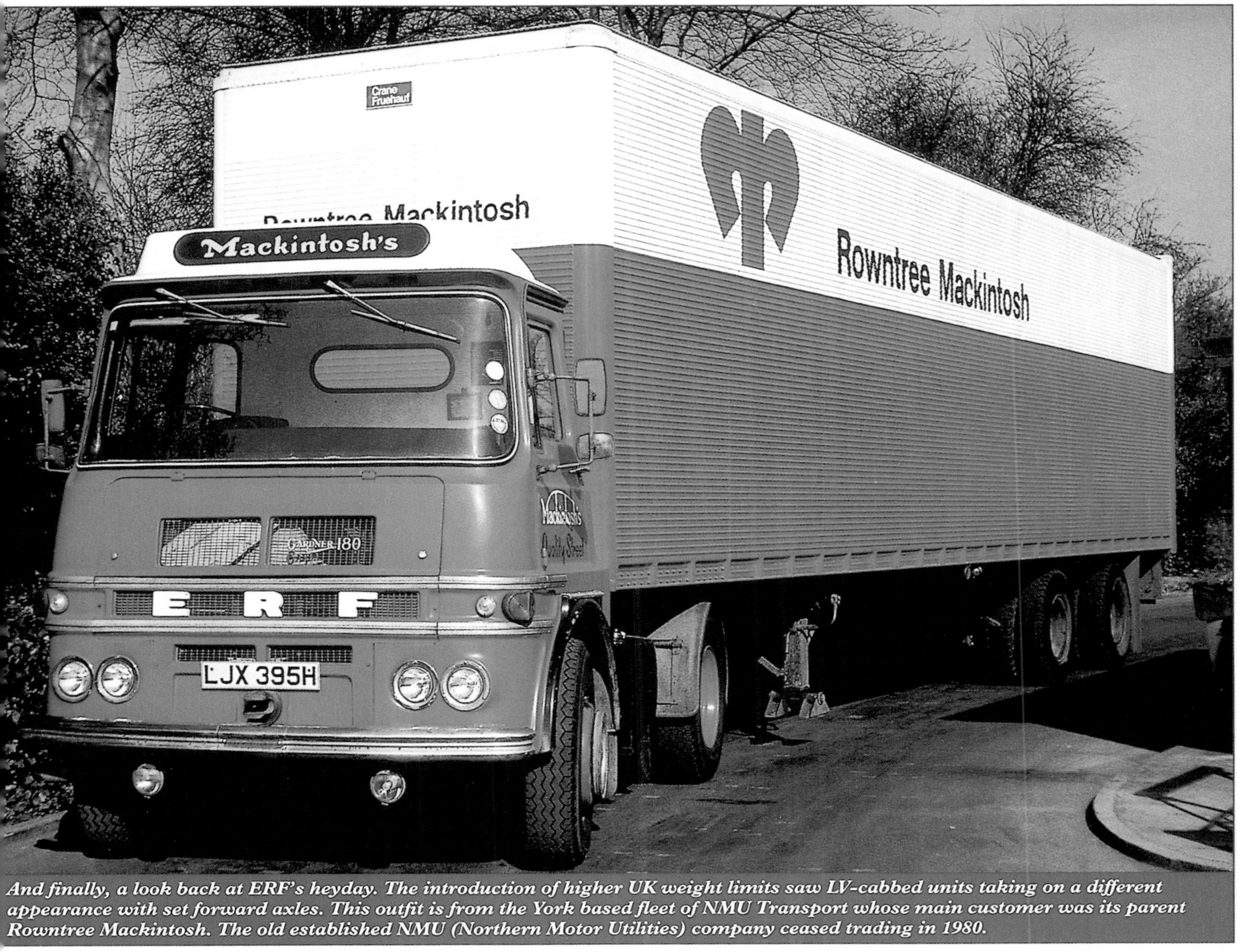

Author

And finally, a look back at ERF's heyday. The introduction of higher UK weight limits saw LV-cabbed units taking on a different appearance with set forward axles. This outfit is from the York based fleet of NMU Transport whose main customer was its parent Rowntree Mackintosh. The old established NMU (Northern Motor Utilities) company ceased trading in 1980.

MIDLAND DISTRIBUTORS
OF COMMERCIAL VEHICLES AND COMPONENTS
Sales & Service Depots

E·R·F

PARRS (LTD.) GARAGE
LEICESTER

'GRAMS: "GARAGE."

'PHONES: 61511/2/3.

C. J. TRIBE & SONS, WISBECH.

G. & B. WATSON, LTD., CAMBS.

E. R. F.

(DISTRIBUTORS FOR EAST MIDLANDS)

GIBBONS BROS., LEICS.

T. GOY & SONS, LINCOLNSHIRE

The above are some of our recent deliveries.

Official **GARDNER** COLD START Service Agents

HEAVY OIL ENGINES

(Photos: Nick Larkin)

FAIRGRO

A selection of fascinating 1970s and 1980s' ERFs in fairground use, including the oldest-known working B-series. Nick Larkin reports.

Oldest B-series?

A much-loved 1975 ERF working hard in fairgrounds is also believed to be the oldest B-series still in service.

Number 23 off the production line, HUT 108N was new to JW Pallett of Leicester as a flatbed. The ERF began its long fairground career with John Brett in the Midlands, who built a ghost train framework on it. Keith Carroll, by coincidence based in the lorry's original 'home' city of Leicester, bought the vehicle and extended the chassis. The B-series' career became even more varied, culminating in it being loaned to a newly opened theme park... in Taiwan!

Back in Blighty, the ERF spent three years at a Cornish attraction before passing to the present owner, Paul Evans, who runs P&LA Evans Family Fun Fairs based at Spennymoor, County Durham.

"It hadn't run for three years before being driven up from Cornwall, but apart from non-active power steering and a blowing exhaust, it made the journey to County Durham without problems," said Paul, who's owned the ERF since 2000.

Since then, the power steering has been sorted out, the Gardner 180 engine (described as 'wonderful' by Paul) has been thoroughly serviced and air leaks repaired. The cab has also been repainted. The David Brown eight-speed gearbox hasn't needed work.

Still hard at work, 1975 ERF B-series.

Paul redesigned the fairground ride, originally constructed by Leslie Vernon Fabrications. The hydraulics were moved by Paul. "Why make work for yourself by maintaining that lot?" he says.

The 35ft long chassis-mounted Funhouse ride consists of several levels and includes a seascape, jungle, enchanted kingdom, Dennis the Menace attraction and a helter-skelter. It's now being extended to incorporate a Scooby Doo theme.

Power for the ride comes from a modern Kubota three-cylinder-engined generator. A wise decision considering that not only do the myriad of fairground lights need powering, but also a sound system and a smoke machine – the latter being rather eclipsed by the 180 Gardner starting from cold!

Paul and wife Lucy Ann take the ERF to more than 60 events across the Showmens Guild Northern Area. Along with the Funhouse, they also have Lucky Numbers Stalls, a Cup and Saucer Ride and an Old Time Springboard. An E-registered

ERFS AT WH

A fine selection of ERFs took over the Market Place at Whittlesey, Cambs, when a fair was held recently.

Looking immaculate, this 1973 locally registered ERF A-series.

Another hard-working A-series looks set for years more work.

UND ERFS

Funhouse – a description which could apply to both the fairground ride and its ERF transport!

ERF cab has been repainted and signwritten. The lorry has a name, of course – Katy.

suitably liveried Ford Transit van is also used for transport.

Paul's had a lifelong love of the fairground, having spent every weekend as a youngster with his Uncle John, who owned waltzers.

"It's a great job, and maybe the only one where you're a mechanic, welder, painter, showman, shopkeeper and spokesman at the many meetings with local councils."

Like many showmen, Paul is keen to dispel the rather tainted image of his profession among some members of the public. "I think because we're going from place to place and living in caravans when we're on the road, some misguided people think we're itinerants. But we have our own house and our business operates under strict guidelines of the Showmen's Guild. Members do a lot of charity work such as giving free rides to local handicapped children. Fairgrounds are also among the longest-established businesses."

Depending on the event, Paul sometimes supplies all the fairground rides and on other occasions, sets up with the owners of other attractions. Some large events have fairgrounds under the name of a particular operator, such as the Sunderland Air Show's fair being run by Noble Amusements.

Paul intends to keep the ERF in service for as long as possible. It's just arrived back from a full service at well-known specialist in the marque, Ritchies of Hetton-le-Hole.

"The attention this ERF attracts is phenomenal, and as far as we know, it's the oldest B-series still in service," he says proudly.

To hire the Funhouse (with ERF attached!) and other fairground attractions, contact P&LA Evans Family Fun Fairs on 01388 777045 or 07836 350507. Any further information on HUT 108N and its history would be most welcome.

TTLESEY

C-series by day, and night!

FAIR THE

A locally owned fairground fleet has a rich assortment of working and preserved classic commercials seen at the recent Stoke in Steam event. David Reed took the pictures.

Fairground attractions

This superb collection of ERFs is owned by locally based showmen Stanworths Fun Fairs and travels to events across the Midlands and Wales.

Three generations of the Stanworth family have been involved with the business, which is run by brothers Henry and Keith. Henry said: "We've always had excellent service from ERFs – and the four we still have are still very reliable and easy to work on."

Star of the fleet must surely be an LV, JRE 790M, which had worked for Dowling Bricks. Originally sporting a Cummins 220 engine, it now has a 250. A cab repaint is part of the work planned for this season.

Then there's RRR 753R, a B-series which ran for many years with Day Lay Eggs; a B-series eight-wheeler AJU 831T was originally a long wheelbase example, but was shortened for its fairground generator carrying duties. C-series A531 HAC is the only fleet member with a Gardner engine.

But if you think this lot's exciting enough, just wait until you hear about the family's restoration project, ERF Sabrina XCD 91! Parts are ready to start the project.

LV is to be smartly repainted for this season.

Above: Despite being the most modern veh with a Gardner engine – the rest are Cumm

Left: B-series was shortened for its present

ERF C (far left), LV and B-series are all hardworking memories of the Stanworths fleet.

RE!

1 HAC is the only member of the fleet

That old dilemma at these events – whether to look at the lorry or the 'official' fairground attraction first.

ALSO SEEN AT STOKE

It was at the Co-op!

Mention the Co-op and everyone who is old enough can remember the 'divi number', even if their memory doesn't go back to the days when 1927 Dennis G 30cwt model, FH 5675, was hard at work delivering coal for the Co-operative Society.

The vehicle was sold in 1947 to a Mr Timms of Gloucester. He kept it until 1961, when it was sold to the present owner Gerald Jenkinson.

"I bought the Dennis for £50, and have had it for 41 years," said Gerald. "I restored it and took it to some of the early rallies that were taking place. It was restored again 10 years ago, when we rebuilt the engine, gearbox, back axle and radiator cores."

Some items relating to FH 5675's time with the Co-op have survived, including a 1941 bill for delivering six bags of coal at £14s 2d (71p). The last job for the Co-op was on October 4, 1946, carrying building materials and timber, the driver on that occasion being Mr N Townsend.

With a period load of sacks, the Dennis certainly looks the part, though it still awaits one small finishing touch. Explained Gerald: "I've got some coal weighing scales, but haven't had time to complete them yet."

Long standing Transit

This beautifully presented Ford Transit pick-up has been with owner Mr A Wilson since 1969, remaining an operational vehicle until 1998.

Bedford pair

Many local connections for Bedford TK, OBF 301W. It was supplied by Bristol St Motors to a Mr Slater, who used it to carry building materials in and around Stoke. It's seen with a beautifully restored Bedford J-type.

SWEET ERF

After more than half a century as a lorry driver, George Wassell now has his own classic ERF, a 1962 LK44G, which brings back sweet memories for many. We meet him.

(Photos: Alec Kermotschuck)

Scrummy confectionery lorry. Did 'Something good from Bury' referred to the vehicle as well as the toffees it once carried?

Windscreen in sections, possibly as there were problems at this stage in making a complete wrap-round screen. The doors were reskinned as part of the restoration.

Move over Willy Wonka – it was Benson's from Bury who really knew how to make sweets. Their superbly liveried red ERFs were a familiar sight for so many years, ferrying precious cargos of chocolate eclairs and toffees throughout the country.

Motorists never forgot seeing a Benson's lorry in their rear-view mirror. They'd see gleaming paintwork and a highly polished ERF oval grille, but instead of the Benson name, there'd be a proud slogan. 'Something Good from Bury' was proclaimed in huge letters, also repeated on the rear of the vehicles.

Generations of children from Godalming to Plymouth must have thought of Bury, Lancs, as a wonderland where the paving slabs were made of toffee and brussels sprouts were banned. You could also imagine cynical older folk from Leeds saying: "So there is SOMETHING good about Bury."

Sadly, Benson's is no more, a Liverpool firm (we think Barker & Dobson) having gobbled it up faster than a quarter pound of best chocolate eclairs some years ago. But now the company's logo is seen over a wide area thanks to a newly restored 1962 ERF LK44G. Spreading the word of 'Something Good from Bury' is the vehicle's owner, George Wassell, someone good from Barnsley who has been involved with lorries for almost half a century.

"We used to see Benson's lorries when I was a kid, and I always remembered that slogan," he explained. "I've always loved ERFs too, and have had a passion for anything old and interesting."

George explained: "In the garden I have a phone box, pillar box, an old park bench, a grindstone and back at the house I've got three horse-drawn ploughs. I don't think they make things today as they used to do."

George is often accompanied at events by grandsons Adam, 16 (pictured), and Jamie, 13.

Slogan extended to the back as well as the front of the vehicle.

Chromework is as immaculate as the rest of the ERF.

Rear doors carefully constructed so you could see the tail-lights when they were open. Two rear number-plates were fitted.

Bensons logo featured a gentleman to rival Sharp's Sir Kreemy Knut, star of our feature on confectionery lorries in the June 2003 issue.

He added: "I'm a very patriotic person, a true Englishman and proud of anything British-built. Hence the Union Jack in the cab of the ERF… and my Union Jack braces!"

George was scarcely out of short trousers when he began work in the haulage industry: "I started as a driver's mate, working for Redfern's Glass Bottles, based at Monk Bretton, Barnsley. I then got called up to do two years' National Service but actually ended up doing three years.

"After that I went straight into driving with soft drinks firm Beckett Brothers of Barnsley, which only had Bedfords."

He then spent some years with Cutts and Fletchers coal haulage of Hoyland Common, who concentrated on Gardner-engined Atkinsons, before graduating to tanker work with George Cooper of Shafton AECs, an AEC stalwart with Mammoth Majors and Mercurys.

After this, George spent 16 years with Bolsover, Derbyshire-based Coalite, then worked for Fearns tankers for 12 years. By then he was an ERF devotee: "I must have spent 25 years on B and C-series ERFs. They were always very reliable and gave sterling service.

"Coalite wouldn't have anything else. Whenever they had one that was different, they always sent someone from ERF to train the drivers. That's how I learned to use the splitters. Some of the drivers used to complain about them, but they were lovely when you got used to them. A lot of drivers were thrown in at the deep end and it wasn't their fault they had difficulties."

After being made redundant, George took a job driving for supermarket Netto: "There was an ERF there and no-one liked it except me." Next came a spell with Londis, then Lanes who carried out work for Next, for whom he now works after the clothing chain took its transport operations in-house. He regularly drives nights, preferring this as there's less traffic. His current steed is a Volvo FM12.

"Things have changed over the years. The worst thing that's happened to lorries is sleeper cabs. It loses a lot of the social side of the job. There's no camaraderie now. People are in so much of a hurry. They'd do a constant 100mph if they could. A trip to Scotland would often mean two or three stops. Now you'd do it in one hit."

He added: "Years ago you got to know the best digs to stay in: George's of Hamilton, Jack Blackburn's place – called something like Almaree or Almaria – in Union Street, Dundee and Big Ethel's at Islington, London.

"There were places like the Stainmore Café on the A66 – I've had to kip on the floors of cafés when the roads have been blocked by snow."

While George appreciates the creature comforts of modern lorries, he says: "I wouldn't have wanted my son to have gone into it, and wouldn't advise anyone else to."

George finds his ERF a great reminder of more relaxed days in the haulage industry. "When I get in it, I feel totally chilled out," he says. The ERF's restoration wasn't so relaxing, however. In fact, it would have raised anyone's blood pressure.

PEN 897 was new to Benson's (note the Bury registration) and was exhibited at Earls Court for the 1962 Commercial Motor Show. The lorry was fitted with a white steering wheel. Bodywork was to typical Benson's spec and built by Jennings – more on that later. A four-cylinder Gardner 4LK engine was fitted.

The ERF settled down to a long career, after which it was sold to showman's use then to preservation, Jack Cook owning the lorry for 15 years and restoring it in the brown and cream livery of Halifax-based George Pickersgill Removals.

"I saw it doing the rounds at rallies for a long time and then noticed the vehicle for sale in the classified section of *CVC*. I went to see it and a deal was done there and then," recalled George.

The lorry ran well, and it was only after getting it home that George discovered just how bad the bodywork was. "All I could see wrong with it were the door pillars, but I had a nasty shock a few days later when I discovered it was rotten from where you sat in the cab downwards.

"I was talking to my son, Sam, and tapping the side of the wagon and there was dust coming out." Investigation revealed major rot to pillars and the cab floor. "If I'd known it was that bad, I don't think I'd have touched it," he said.

The only option was to continue.

George removed the front panels to find this amount of rot in the timber frame. At least there was enough left to make a pattern.

Taking shape as the new timber frame sections, this time on the offside, are put in place. Note the inner wing being tried here for size. The floor was the next job.

Not an easy task – refitting the curved glass. Note the glazing strip on the corner joining the two pieces of glass together, held in place by a rubber seal.

ERF still in Pickersgill livery and about to depart for the paintshop.

Panels sorted and topcoat of Mason's Red applied. Next, the signwriting!

"I sat down and listed all the parts needed, and it was a long list. The ash timber frame over both wheels was rotten, as were the aluminium front wrap around panels and the lower door panels.

"A lot of work was needed here and it called for skilled assistance. I admit it was beyond my capabilities and I was less than pleased with the whole affair to be honest," George remembered.

"Being a humble lorry driver, all I could do was to finance the job by working weekends, paying craftsmen for their time and material as it went along. The whole of the cab floor on one side and half of the other needed replacing in wood."

Dave Sykes, a skilled joiner and cabinet-maker friend, dealt with the framework, painstakingly removing rotten existing sections and where possible using them as patterns, then putting on the front panels.

The ERF went to Stuart Bebbington of B&C Vintage (01606 553800/07855 682824). He started as an apprentice with bodymakers' Jennings and worked on several Bensons' vehicles, probably including this one. "They had four and six-wheelers with these bodies, which were virtually to their own spec," Stuart recalls. "I've done many cabs on classic vehicles, but it was strange to think I probably worked on the building of this one 40 or so years ago."

Stuart reskinned the ERF's front and rear doors along with further panelwork also being needed. Classic commercial enthusiast and owner of an Austin K2 recovery lorry, Glyn Heseltine of Barnsley, fitted a replacement windscreen. This is in four sections with a corner connecting strip at each side to hold it all together.

The ERF was then taken to master painter and signwriter, Jack Rowley of Crewe, who rubbed down the outer panels where necessary and after priming applied original-specification Mason's Red paint, keeping faithful to the original build sheet which George has. With the help of a photo, he even recreated the original signwriting – a wonderful job!

Electrical guru Trevor Barlow, whose AEC Mercury and Foden S39 were featured in last month's *CVC*, worked his magic on the ERF, including reactivating the speedometer.

As for the engine? "It needed nothing apart from a damned good service," said George. "The same went for the brakes, it'll stop on the handbrake."

Final finishing still has to be done, with the dashboard needing to be re-covered and the seats receiving some attention.

George is now exhibiting the ERF at shows over a wide area, often accompanied by his grandsons, Adam, 16, and Jamie, 13.

The vehicle had already been fitted out as living accommodation. George was given a new cooker and has rigged up some bunk beds (using old grab rails from buses supplied by his friend, bus dealer/breaker Alan Hardwick), so now there really is a family home from home. "There's even a radio and a telly, so we're short of nowt."

George has a Benson's liveried bike, believed ex-Post Office, to go with the lorry. "It's got to be about 50 years old and complete with the original leather-sprung saddle. My family bought me two new wheels and tyres for my birthday. I ride it, but once I'm on it, it's a problem as I'm too short in the leg to reach the ground, which has resulted in me nearly falling off a time or two."

Final comment on the vehicle from George has to go down as one of the greatest-ever comments on the subject. Was, we asked him, the finished lorry worth the effort and money put into it during the restoration?

"Oh aye, it's like having another kid, isn't it?"

APPEAL

If you have any photos of Bensons' vehicles or memorabilia, George would love to hear from you on 01226 752855, preferably 10am-3pm. Remember, he works nights!

'Something Good from Bury' owned by 'someone good from Barnsley'. George with his trusty Benson's-liveried bike.

White steering wheel is instantly noticeable. This is believed to have been fitted for PEN 897's star appearance at the 1962 Commercial Motor Show – or some say so the vehicle could be seen more easily at night.

PEN 897 when new. Note how accurately the signwriting has been recreated. More Bensons nostalgia next month.

ERF gets attention on the road wherever it goes.

GEORGE'S TIPS FOR RESTORERS

1. Have a good look around the vehicle, including the underneath, for any signs of rotten timbers and panels. There could be hidden horrors, even if the vendor honestly doesn't know they're there.
2. Curved windscreen glass costs a fortune to have made so bear this in mind when looking at a vehicle that needs some work.
3. Employ the skills of specialist tradespeople if there are jobs you can't do yourself, but remember, their labour and the materials aren't cheap.

Bensons ERFs

Last month's major restoration feature of George Wassell's 1962 former Bensons confectionery ERF LK44G has prompted Carl Johnson of REVS to send us these archive photos of lorries built in Sandbach extolling: 'Something good from Bury.'

Now restored LK44G, PEN 897, seen when new in this Jennings coachbulders publicity shot, and not looking that different from today!

Something good from Bury 1950-style, again with Jennings body finished to an extremely high standard.

Back to 1949 with an ERF LK44. Note the early flowering of Bensons' toffee mascot seen on the door.

The 1960s now, and Bensons Toffee has become Bensons Quality Sweets with this LKG44, as depicted in this ERF publicity shot.

SANDBACH

ERF celebrated its 70th anniversary in 2003. Martin Perry looks at some fascinating still-in-service survivors.

ERF celebrated its 70th anniversary in 2003, having built more than 90,000 lorries since Edwin Richard Foden and Ernest Sheratt first set up shop in Sandbach.

There have been many 'firsts' in ERF's history, starting with the famous chassis number 63 in 1933 through to the first Chinese Six, the ground-breaking Kleer Vue KV range (the first production vehicle with curved safety glass), developed with Triplex.

From 1959, ERF was the first British truck maker to offer Cummins engines in production vehicles and later pioneered the air-actuated parking brake. Both developments stemmed from Sheratt's forward-thinking policy of studying the latest developments in the US trucking industry.

It has been suggested that at least a third of the 90,000 lorries that rolled off the production line at Sun Works since 1933 are still in existence. The famous SP cab, combining rust-free and very strong fibreglass on a sturdy steel frame, accounts for many of these. They are popular with travelling showmen who require reliable lorries capable of standing out in all weathers, free from the dreaded tin worm that proved the downfall of many competitors.

Here are some other survivors.

Note: If you'd like to learn more of ERF, look up REVS (the ERF Society) on its excellent website, www.erfhistoricvehicles.co.uk

The popularity of ERF on the British fairgr
wheeler. New to cider makers Showering o
by the De Vey family, whose fairs are a vita

With ERF claiming to have built the first twin-steer Chinese Six, it is fitting that I include an example of the rare configuration. And what better than BRF 590T, WR Munro's 1979 Gardner 180-engined 66TS B-series, seen plodding steadily through Bromyard in October 2003, complete with a roped load of roofing sheets – every inch a proper lorry!

Introduced in 1982, the C-series was one of ERF's greatest successes, and a good number remain at work. This classic eight-wheeler powder tanker, YEW 444Y, is operated by Walsh's of Crewe and seen on pounding up the A49 near Ludlow in May 2003.

Just a couple of years newer is this smart A-series, part of the excellent fleet operated by Jolly's Circus in the Welsh Marches. Photographed in September 2003, it was new to famous Yorkshire carrier Hargreaves as a tractor unit, and is a model 64 CU with Cummins power.

Old and new on the fairground, but both ar
Dorset steam fair are 928 HFU; immaculate
years its senior! Gardner-powered FUY 690,
remains hard at work at 32 years, carrying

SURVIVORS

*ied by this superb model 68 GXB eight-
:, GYA 827K is kept in immaculate shape
Country showland scene.*

*ıssic tradition. Seen here at the 2002 Great
ping company with an LV unit some 22
famous Evesham fleet of Marshalls and
vintage fairground stuff.*

Above: The M-series 16-tonner was in production from 1978-82, but just a handful survive today. One is this Gardner-powered flat, UJA 249X, which started life in October 1981 with the East Lancs Chemical Co and today works in Worcestershire with CJ Hughes of Twyford. It's seen here trundling south through Ross-on-Wye.

Left: One of the oldest ERFs still in use is Emerton Roofings' LV four-wheeler, which Gerald Emerton collected from Sun Works in 1969. It remains a vital part of his Nantwich business over 34 years later. HMA 918G is a rare 64P model fitted with a Perkins 6.354 engine.

Another C-series, this excellent 20-year-old Gardner-engined six-wheel tipper still carries quarry products in Herefordshire with Tom Bowkett of Painscastle. Here she is with another load of limestone at the Dolyhir Quarry, Kingston.

HEMTT
REVIEW
OSHKOSH TAKES BRITISH TANKER CONTRACT
IWM North
NEW GM PICK-UP TRUCK
BLAST FROM THE PAST

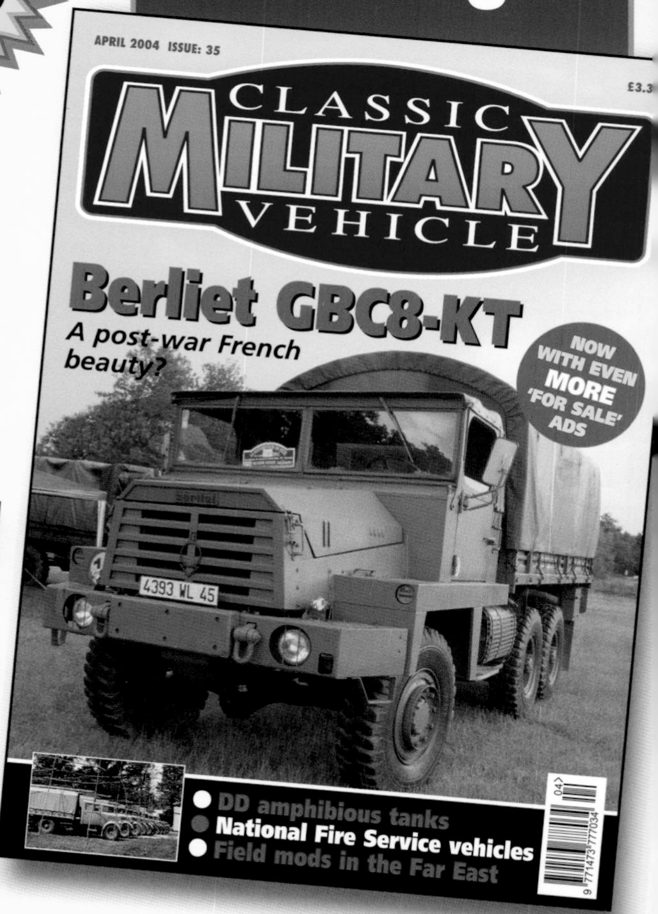

APRIL 2004 ISSUE: 35
CLASSIC MILITARY VEHICLE
Berliet GBC8-KT
A post-war French beauty?
NOW WITH EVEN MORE 'FOR SALE' ADS
4393 WL 45
DD amphibious tanks
National Fire Service vehicles
Field mods in the Far East